Through the Crust of the EARTH

Through the Crust of the EARTH

LORD ENERGLYN

Macdonald·London

Copyright © Lord Energlyn 1973
First published in Great Britain in 1973 by
Macdonald & Company (Publishers) Limited
St Giles House, 49–50 Poland Street, London W1

ISBN 0 356 04604 4

Made by Roxby Press Productions
55 Conduit Street   London W1R ONY
Editor: Tony Aspler
Picture Research: Penny Brown
Design and art direction: Dodd and Dodd
Printed in Great Britain by Morrison & Gibb Limited
London and Edinburgh

Chapter One

MYSTERIES OF THE UNDERWORLD

Man has always had an ambiguous relationship with the interior of the earth. From the earliest times he has feared the underworld, casting it in religions as an abode of evil and the ultimate repository of the damned. But at the same time he found beauty in the earth – gold, silver and precious stones – and minerals essential to his life on earth – coal for fire and ores for metals. So man had to put aside his fears and descend into the earth. With this descent came knowledge of the workings of the world, and for the first time man learned how minute examination of rocks could explain many hitherto baffling aspects of his past and give a guideline to the nature of his future.

The concept of Hell in the Judaeo-Christian tradition evolved from pagan mythology. The word Hell is derived from the Anglo-Saxon, probably from 'hylja' – an old Norse word meaning 'to conceal' or 'to cover'. This suggested that it was originally thought to be concealed in the torrid region near the centre of the earth. In traditional Norse mythology the goddess of the infernal regions was called Hel and she ruled over Niflheim – an underworld of cold and darkness divided into nine regions.

In Greek mythology the kingdom of the dead was ruled over by the god Hades (or Pluto) and his queen Persephone. According to Virgil it was located beneath the secret places of the earth. The ancient Greeks divided the underworld into two divisions: Tartarus, the lower region, and Erebus where the dead go as soon as they die. Virgil tells us that the Inferno is encompassed by five subterranean rivers, Acheron, Cocytus, Styx, Phlegethon and Lethe and that the aged boat-man Charon ferried the dead souls across the one called Acheron, the river of woe, to Tartarus which was the gate of Hell. This was guarded by the fiercesome three-headed dog, Cerberus. From this topography it seems that the Greeks regarded the underworld as a gloomy subterranean realm or a distant island which no man-made ship could reach.

The early Jews used the term 'Sheol' as the place of the dead. Sheol meant a vast cavern under the earth – a huge tomb where all the spirits of men were gathered. By the time of Christ the Jews had acquired the belief that wicked souls were punished after death in Gehenna – which took its name from the Valley of Hinnom, outside Jerusalem where fires were continually burning to consume refuse. The Muslims described Hell as a hierarchy of concentric circles in the shape of a crater above which is a bridge as narrow as a sword edge. To negotiate it successfully meant a life in Paradise; to fall ensured eternal damnation.

The Hindus believed in twenty-one Hells beneath the nether world and the Buddhists say that in the realm of Kamacavara there were eight hot Hells and eight cold ones. They also maintain that the god Brahma ruled the world and that during his waking hours he created and then destroyed his creatures in his sleep. Such actions, to the Hindu mind, were the origin of fire, floods and other catastrophes. Similarly, the Greek philosopher, Heraclitus, in the year 500 BC was teaching his pupils that 'the world is and will be eternally a live fire, regularly flaring up and dying down again'. These concepts were clearly the product of direct experience of the storms at sea, destruction by earthquakes, land-slides and avalanches of mountainous masses and the eruption of volcanoes.

Throughout the ages then, theologians, poets and philosophers agreed generally as to the location of Hell: it was a 'bottomless pit' in the centre of the earth as far away from Heaven as geographically possible. Although a certain divine, Tobias Swinden disputed the nature and location of Hell in 1714; he put forward a case for the sun as the site of Hell since its size enabled it to accommodate the enormous number of the damned souls and their resurrected bodies, as well as being permanently fiery. The centre of the earth, he argued, could not contain sufficient combustible material to provide eternal fire and brimstone, nor could it burn without air. Hell, in spite of Tobias Swinden, was literally and metaphorically at the centre of the earth; this concept dominated the thinking of man right up to the seventeenth century.

If man in classical times believed that the gods who controlled the elements lived under the sea or in the earth then their helpers by definition must live there too. The idea that gremlins, elves or fairies were responsible for crop failure, cattle disease and suchlike misfortunes was very much a part of popular superstition up to our own times. Fairies, now those benign and ethereal creatures of children's books and Christmas tree decorations, were to countryfolk mischief-makers who were to be feared. Such creatures lived underground out of sight of human eyes, usually in the roots of trees. Fairyland was believed to be an underground region and bore a great resemblance to the kingdom of the dead in folklore tradition.

Myths and Reality

It is not unreasonable to interpret some of these myths in terms of our modern knowledge of the earth and natural events. For instance, Indian cosmologists 6,000 years ago maintained that the earth consisted of seven continents spreading out like lotus leaves. Today we know that continental masses are distributed as masses riding on a molten interior. These are now called continental plates which have drifted apart over millions of years to arrive at their present locations on the globe. We now have proof that large areas have been fractured by faults. It may well be that the Greeks had come to much the same conclusion when they said that the Rock of Gibraltar and the rock mass of Jebel Musa across the strait had been torn apart – even if they did ascribe the feat to Hercules!

The legend of Atlantis – a vast mythical island believed by the ancients to have existed in the Atlantic – finds its place in our modern interpretation of the elevation and subsidence of a continental mass. It was first mentioned by Plato in about 400 BC. His great-great grandfather, the Athenian sage Solon, was told of it by Egyptian priests who claimed that the King of Atlantis had attacked the Athenians but was defeated and that soon afterwards Atlantis, with all its inhabitants, was drowned as the continent subsided into the sea. It is probably coincidental that the Assyrians and the Babylonians begin their chronology with events which took place about 11,500 years ago and that the modern dating of the origin of the Gulfstream, along the supposed location of Atlantis, was about the same time. This is yet another example of how legend and fact so frequently coincide.

From Stone Age to modern times myths such as Atlantis have survived because many of them are based on observation – the evidence of man's eyes or his reaction to his environment. Certainly this has been proven in medical science: the fairy rings of toadstools from which the witches took the soil to make their healing brews have now been shown to contain antifungal substances which behave like antibiotics. Innumerable examples of this kind could be cited and through them all runs the thread of man's intense response to his environment.

Man the Stone-maker

Man's proven existence on earth has spanned at least $1\frac{1}{2}$ million years, but this is only one six-hundredth part of the time it has taken for him to come down from the trees. The moment he left the sanctuary of the trees for life on the ground, he depended on his wits for survival. Other animals more fleet-footed and ferocious would soon have disposed of man's presence on the land had he not been able to create weapons for defence and attack. So stone implements became early man's most precious possession. Stones, especially sharp-edged ones, had to be at hand to hurl at animals, and these he found in river valleys, gorges and caves. So the siting of man's first centres of habitation was dependent primarily on defensive factors – a ready supply of stones. It is therefore not surprising that Dr Leakey and his wife discovered the oldest known remains of man in the now famous Olduvai Gorge in Tanzania. Labouring in intense heat,

The Bridge over Chaos: a mythical concept of the underworld as a place for the dead.

these two dedicated archeologists ignored much hardship to unearth the story of ancestral man from the relics of bones and implements buried in the sands and gravels of the gorge.

The Leakeys established that this primitive race of humans took up residence there using the sharp volcanic pebbles as tools and weapons to provide the basis of their carnivorous existence. The remains of their bones have revealed that this community of ape-men settled in the Olduvai Gorge at least 1½ million years ago. As one would expect their implements were crude – nothing more than pebbles split to expose sharp edges which they used to kill and carve meat from the bones and hides of their victims.

The records in these rocks show the community becoming in-

creasingly efficient in trimming and shaping one piece of stone with another. Sparks must have flown as stone struck stone, yet those primitive artizans never realized that they were exercizing the one ability that was destined to make man unique among animals – the ability to make and control fire. Only certain types of stone would behave in this way. Thus man found by trial and error that rocks which would splinter like glass were the ones he needed. Relatively few rocks would do this; and to own flint or volcanic glass was to be a wealthy man in Stone Age society.

With two pieces of rocks like flint early man developed the skills of a precision engineer: he chipped out axe-heads, knives and arrow-heads perfectly balanced and highly polished. Even bracelets and jewellery were fashioned in this way by discovering how to knock a hole through flint.

The First Miners

But valuable rocks could not always be had just by picking them up from the ground. Somewhere man had to start digging into the earth to get at his raw materials. The only available tools for ex-cavation were bones – deer antlers and shoulder-blade bones – to act as picks and shovels. With such tools these primitive miners achieved remarkable successes. A spectacular example is the site known as Grimes Graves in Britain which constitute one of the best known groups of Neolithic flint mines. They lie in the gentle slopes of a dry valley in south-west Norfolk. The Neolithic flint-miners had learned that large nodules of flint occur naturally as beds in chalk, and that certain layers of flints were more suitable than others for making tools. One particular level at which these flints occur in the chalk was known as the 'floorstone'. To reach it, Neolithic man sank shafts over 30 feet deep with only their deer-antler picks.

From numerous pits and their radiating galleries, the miners fol-lowed the thin band of floorstone flint. In all, no less than 306 shafts were sunk and linked up underground within an area of about 34 acres. It is not certain how the miners managed to lower themselves to the bottom of the shaft or how they hauled the excavated flint waste chalk to the surface. In the more shallow shafts traces of steps are identifiable but in the deeper holes it seems likely that they had invented rope-ladders made of plaited thongs of hide and wooden treads. In some places there are signs that a horizontal beam or tree trunk was placed across the shaft. This suggests that bags made of hide were hauled up across these beams. The skill required for this

The standing stone of North Ronaldshay (side and front view) erected by the ancients either as a landmark or for some mysterious ritual.

(left) A triple 80 million volts flash of lightning – the power of nature unleashed. The lightning fork is striking a tree.

The Ring of Brodgar in the Shetland Isles, symbolic of the unwritten history of man's interest in the hereafter.

Bats at rest. Thousands cling in an orderly fashion to the roof of the Carlsbad Cave, New Mexico.

intricate operation singles out these men as the forefathers of mining.

Grimes Graves is but one example of Neolithic flint mines; at Apiennes in France shafts over 30 feet deep revealed several layers of flint which were excavated by Neolithic man with antlers – picks and bare hands. In a flint mine at Obourg in Belgium, a skeleton of a Neolithic miner has been discovered buried by a fall of roof and he was still holding his antler pick! Few tragedies of this kind have been unearthed which is quite remarkable considering that the Neolithic miner burrowed ahead without using timber supports and appeared to disregard the need for adequate ventilation. He needed light though and for this he used molluscan shells containing animal oil. With these dim, flickering 'chalk lamps' the air must have been foul and yet his productivity as a flint miner was incredible.

Underground Burial

Whichever aspect of prehistory is studied, the basis of man's development is his deeper and deeper involvement in the products of the earth's crust. But the records of these activities are only discovered by chance and even then they are subject to various forms of interpretation. This is unfortunate as the Stone Age era of man's existence on earth covers nine-tenths of human antiquity. Yet it is possible from the discovery of skulls and their association with cave art to establish another preoccupation of early man with the earth – its use as a place to dispose of his dead. The Neolithic people of Grimes Graves had a casual attitude to death; they were not above using human femurs as tools and they were somewhat haphazard in the way they dispensed with corpses down discarded mine shafts. But we know with reasonable certainty that early man soon developed a religious outlook on death.

Perhaps the oldest example of ordained burial is that discovered in a cave at Chon Kon Tien near Peking. In these cave deposits, which go back at least half a million years, skeletal remains of six children, two adolescents and twelve adults were found in positions which leave little doubt that they were carefully interred after death. Moreover, only the skulls had been buried or preserved. All the evidence suggests that these cave-men carried the corpses outside and when the flesh had disappeared, they carefully recovered the skulls for burial.

There is widespread evidence in Europe, the Middle East, Africa and Indonesia of what has been described as the cult of human skulls. At some stages in the development of this cult, strange rituals existed involving animal skulls. For example, in a cave on the Drachenloch in Switzerland, six rectangular stone boxes were found containing skulls of the cave bear. Tightly packed, each skull was facing in the same direction. Many similar instances of ritual have been discovered in caves and in open burial grounds, including the orderly heaps of mammoth bones found at Hanci in the Ukraine. The reasons for this form of ritualistic burial are unknown but some

Grime's Graves, Norfolk, England, one of the earliest mines on earth. Stone Age man delved with bone picks and bare hands thirty feet into the chalk to find and remove beds of flint.

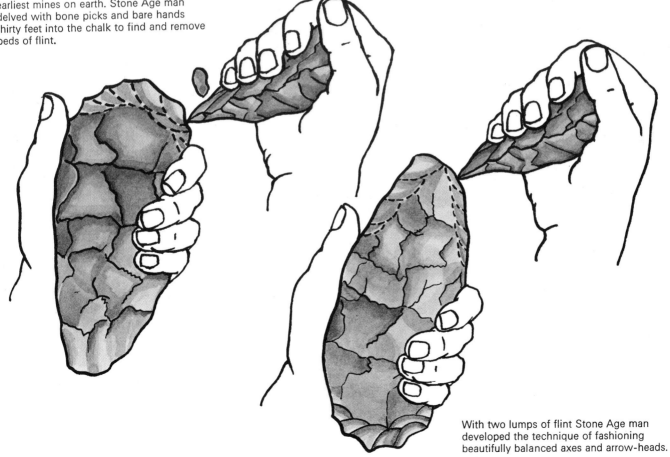

With two lumps of flint Stone Age man developed the technique of fashioning beautifully balanced axes and arrow-heads.

Art museum Stone Age style. The artefacts
of primitive man decorate the walls of the
Adjunta Caves, India.

anthropologists have suggested that they resemble the practice of the Alaskan Eskimoes who hide unbroken bones of their dead animals under stones to aid the process of reincarnation.

At one time few people believed in Stone Age graves and funeral rites, but the evidence is now too strong to deny their existence. The Neanderthal Man found in the deposits of a cave near Dusseldorf, Germany, must have been buried there; otherwise his corpse would have been devoured by hyenas. Moreover, when two skeletons of Neanderthal Man were found side by side in a cave at Spy in Belgium, the only prognosis could be that these privileged people were deliberately and ritualistically buried by their fellow men. If such was the case, then we can safely infer that leadership or sovereignty was as natural an order of affairs then as it has been throughout recorded time. We see here unmistakable signs of man's incipient belief in the supernatural and the hopes for an after-life. Such beliefs are inextricably bound up with the idea of an underworld beneath the crust of the earth.

Exploring Caves

Fear of the unknown has been a curious and yet irresistible force which has impelled man to accept dangerous hazards in search of knowledge or material gain. There seems to be no logical reason why such great importance should be attached to the acquisition of fame or wealth, but no risk seems to be too great to deter some from the pursuit of Mammon. Why should we regard a man as wealthy just because he has a large quantity of gold or diamonds? Why should a man become famous for climbing the highest mountains? In this day and age is there any sound reason why we should face self extermination for the sake of some ideology or be tempted to unleash the chain reactions of nuclear energy in order to destroy some offensive aspect of life on earth?

None of these unpredictable aspects of human behaviour make sense when we consider the rewards which could result from a positive reaction to the human environment. But then man is only one unit of life on the fabric of this planet. Too frequently we flatter ourselves by assuming that we are the dominant factor on earth; but when ranged alongside the forces of nature our command over sources of energy is comparatively feeble. Even the force of an atomic explosion is exceeded a hundred-fold by few feet of movement of the solid rocks of the earth's crust. Nevertheless, these catastrophic releases of energy fail to chasten us or lessen our desire to conquer the forces which have formed this planet and endowed it with that unique characteristic we call 'living matter'. This struggle of mind over matter is not the exclusive prerogative of *Homo sapiens*. Every aspect of living matter is engaged in such battles – a fight for survival against other living forms impelled by some urge to travel the road to supremacy.

A chalk goddess found in Grime's Graves, Norfolk, England.

'Man is only a reed,' the French philosopher Pascal has written, 'the weakest thing in nature; but he is a thinking reed.' And thinking involves curiosity. Man sees a hole in the ground and his immediate reaction is to enter it – why? Whether he does or he does not is probably conditioned by the balance between his fear and his curiosity. This dilemma must have been as strong a motivation of the actions of Stone Age man as it is today.

Just as some men have an irresistible urge to climb mountains, others cannot resist exploring caves excavated by nature into the crust of the earth. It would be idle to assume that the average person is interested in the geological reasons for the existence of a hole in the ground, but, once inside a cave, his mind is impressed by the nature of the rock formations he encounters. He is also soon made aware of the fact that he is not the first form of living matter to enter this realm, and this chastening thought in itself stimulates his curiosity. The cave is a remarkable sanctuary for all kinds of wild life. The lacy ferns and clumps of moss which usually grow luxuriantly around the entrance soon disappear as the light is cut off. Further into the cave, the environment is suitable for bats and they breed tremendous colonies. In the huge caves of Kentucky, thousands of grey bats cover the ceilings of the caves where they distribute themselves meticulously into regular patterns. Animals will even adapt themselves biologically in order to live in this subterranean sanctuary. The snail Oxychilus normally eats nothing but plants outside a cave, but inside it adjusts its metabolism to be able to consume bat droppings.

The psychological effects of delving into caverns is a fascinating aspect of man's behaviour. Some people are unaccountably terrified by the claustrophobic atmosphere of a cavern. And even hardened cave explorers find that after a time they are unable to escape the sense of oppression caused by what may be described as the resentment of rocks at this violation of their geological privacy. Solid masses of rock may suddenly erupt during such intrusions. The mere movement of a body can destroy the delicate equilibrium of atmospheric forces which has up until then held the rock face in position. The resonance of a piercing noise has been known to cause cave walls to shed lumps of rock in the most alarming fashion. The physical and geological reasons for these events are now fairly well understood. Blocks of rock are usually poised precariously as a result of the solvent action of water so that any slight change of atmospheric pressure caused by opening up a cave will cause collapse.

The discovery of a new cave always arouses great interest and especially if it is found to contain human remains. These caves are always more exciting than man-made mines as they are holes dissolved or eroded into the ground which dramatically change in shape and size. Rain water seeping into the ground becomes acidified by carbon dioxide. Acting as a solvent, particularly on limestones, this acid percolates through cracks and converts the calcium carbonate of the limestones into a soluble bicarbonate. In this way cracks are

An eighteenth-century engraving of the
activities of the early miners.

opened out into caverns and huge caves, the size and shape of which
is controlled by the nature of the limestones – the more soluble parts
being corroded first. Soon the solvent waters become overcharged
with calcium bicarbonate which begins to precipitate the lime, drop
by drop, to form icicles of stone called stalactites and stalagmites.
When these join miraculous multi-coloured pillars and pavements of
lime minerals are created. But this simple chemical phenomenon is
complicated by the almost capricious changes in the solubility of the
limestones and the pattern of cracks for the water to enter. The water
must have an outlet and this may be many miles away from the
point of entry. As a result, caverns many miles long are not un-
common and they often contain underground rivers, lakes and even
waterfalls. To the explorer, these are a source of adventure since they
present the type of hazard his compulsive nature demands.

Practically every state in America contains a cave because limestone is a widespread deposit. Many of them have been found by
accident. The fabulous caves of Carlsbad were found this way by a
cowboy named Jim White. Riding from Texas into New Mexico
he saw what appeared to be a cloud of smoke rising in the distance.
He rode towards it and found to his astonishment that it was literally
a cloud of hundreds of bats flying out of the entrance to a cave.
Putting fear aside he entered and discovered one of the largest underground chambers in the world, adorned with every conceivable type
of stalactite and stalagmite in an exotic range of colours. He had
entered a chamber 1,800 feet long which arched to 255 feet. It was in
this cathedral-like ceiling that the bats roosted.

Mammoth Cave in Kentucky was discovered in much the same
way. The limestone plateau there is pierced by over 60,000 swallow
holes dissolved from the rock within connected by over 30 miles of
continuous caves and passages.

In Austria a huge block of limestone called the Karst is hollowed
out like a sponge. As early as 1748, Francis I, the Emperor of Austria
ordered Nagel the mathematician to examine these caves, which he
did, and succeeded in descending to a depth of 400 feet into one
called the Mazocha Chasm. In the Middle Ages the caves of Lam-
pecht Sofenloch were vigorously searched for the legendary riches of
the White Lady which were supposed to be guarded by a monstrous
black dog. So many failed to return that it was decided in 1703 to
wall up the entrance to prevent the black dog from claiming more
victims. In the Dgurdgura labyrinth in the Kabylie district of Algeria
the mummified body of a shepherd was discovered by an American
tourist in 1923. The victim is alleged to have been in search of hidden
gold but was trapped in the mummifying atmosphere of the cave.

Water and Life

Whether it be the formation of caves by natural processes or by excavating mines, man always encounters water when he enters into
the earth. Water is one of Nature's most capricious substances. As

Caves, like arteries dissolved by water,
thread their way through limestones and as
they dry out leave behind deposits of lime as
stalactites and stalagmites.

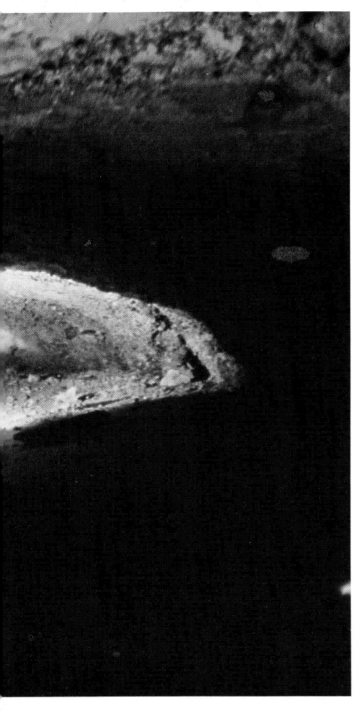

steam it behaves as a gas. As ice it performs as a solid. In all its conditions – liquid, solid, gaseous – it remains simply H_2O, but its physical properties are vastly different. Without water plants and animals cannot survive or even originate. As ice the degree of expansion it creates splits rocks; and as moving glaciers it sculpts the land to shapes unequalled by any other natural eroding agent. In the bowels of the earth the heat converts water into steam which becomes one of the main compulsive forces in volcanic eruptions.

Was it a pure accident that our spinning planet acquired an atmosphere containing water? If so, the origin of life was the result of this accident because without water living matter would neither form nor survive. It is possible that the moon had a moist atmosphere at some point in its history, but it is now clear that it did not last as no living organisms have yet been found. A mechanism for the formation of water is provided by the simple experiment of burning hydrogen in oxygen. Since water is easily formed in this way, might it not be that the red hot particles of cosmic matter which whirled and combined in space to form earth, provided the source of ignition for these two common gases to create molecules of water? If such a hypothesis cannot be proved at least we know that water can be created within the crust of the earth since enormous volumes of water are expelled into the sky by volcanoes.

Once water had been formed, clouds would have appeared in the sky. Within these clouds vast quantities of electro-chemical energy would have developed capable of being released as millions of volts in flashes of lightning. Such discharges would have been quite capable of causing carbon, hydrogen and nitrogen to combine to form compounds called amino-acids. These are the nitrogen-rich organic acids without which no living matter can survive. A few years ago, support for this idea is given by the brilliant experiment carried out by two American scientists, Urey and Miller who subjected carbon and nitrogen to a high voltage discharge similar to lightning. Inside their flask amino-acids formed in this synthetic way. Thus, in claiming that this demonstration has provided the clue to the origin of living matter the role of water is paramount. Without moisture in the sky to form clouds there would have been no lightning to create amino-acids, and with no rain these molecules would have been denied the chance to survive and reproduce other organic molecules to form living tissue. Similarly, without water there would have been no oceans and the eroded land surfaces as we know them on earth.

Living Rocks

Water is essential in the formation of rocks and the production of new rocks from old ones; just like living matter their seemingly inorganic nature is equally vigorous in such reproductive processes. Because our planet has a plentiful supply of water it is unique in the solar system in that it can produce its rocks and the living matter these rocks sustain.

Just as atoms of carbon, hydrogen and nitrogen and oxygen combine to yield molecules which create life, so do atoms and molecules of metals and silica combine to create rocks. In the reproductive sense, one rock will give birth to another and this is the way in which the present land masses have been developed and will go on developing. By studying these rocks we can trace the history of this singular and restless planet for over 4,000 million years and show that the materialism of this environment has led to the creation of mankind.

Outcrops of rock provide the first glimpse we have of what lies within the crust. Linking up the evidence which these rocks possess has become one of man's most rewarding mental exercises. Before the discovery of scientific principles, discerning eyes were able to piece together these scraps of evidence and with swift, intuitive genius, men dug holes in the ground at exactly the right points where valuable concentrations of minerals occurred. Through hindsight it is possible to give sound geological reasons for these remarkable feats of prospecting but at the time the miners had no prior knowledge to work with. These early prospectors laid the foundation of our growing knowledge of the contents and geometry of the earth's crust. It would be flattering to suggest that such men dug in the ground out of a compelling need to find out what was under the ground beneath their feet. They were of course after the metal-bearing ores and the gems in the rocks but the knowledge they gained through experience – the biproduct of their labours – has enriched man's understanding of his environment.

As a consequence of our dependence on the elements, mankind is inextricably bound up in geological history and the future of the crust of the earth. As a product of primitive biological molecules mankind has inherited by evolution characteristics which have spanned at least a thousand million years. The most remarkable is the ability to survive. Throughout the whole pattern of biological evolution we see new organisms designed to replace others; however cruel this process might appear, this was how mankind emerged to dominate life on earth.

Survival of the fittest does not always imply physical superiority as men and women could not be described as such. The superiority of *Homo sapiens* stems not from physique but his highly coordinated nervous system centred in the brain. It is by brain power that mankind has achieved supremacy. But it does not end there; each brain develops in an individualistic manner, so that one person's mental processes and reflexes differ from another's and make the one appear intellectually superior to the other.

We must not lose sight that in addition to the brain and its rational processes man also has instincts and desires which express themselves in such compulsive behaviour patterns as greed and ambition. Such drives are not exclusive to man, however. We inherited these characteristics from ancestors who belonged to the lower orders of the animal kingdom. We tend to think of serfdom for instance as a political idea; it is nothing more than the 'herd instinct' which characterizes the whole of the animal world. In short, like all other animals, *Homo sapiens* is basically a materialist, but for some mystic reason or another he strives hard to attach spiritual motives to his way of life. Unlike the true apes, for example, he is prepared to share his food and, within limits, his material possessions. But when greed becomes predominant, individuals of exceptional brain-power acquire immense power over the material possessions and the lives of the 'common herd'. Consequently, no matter how one looks at life on the crust of the earth, man is inextricably involved in its atmosphere, its soil and the underlying rocks and minerals as raw materials for his very existence.

The Great Hole of Kimberley, South Africa. The blue soils of this vast mine are rich in diamonds.

Chapter Two

THE RICHES UNDERFOOT

The quest for gold in the bowels of the earth. This shaft was sunk two miles into the ground of a South African mine.

The Noble Metal

One of the world's biggest gambles is the hunt for minerals. The rewards are high, but the odds are long and most experts agree that the major finds have been more a matter of luck than scientific judgement. Stone Age man was just as likely to strike it rich with flint as our grandfathers were with gold: although flint soon lost its status as the index of mineral wealth.

When man discovered the control and use of fire he applied it to the smelting of metals. But the first metal he used was gold which needed no smelting process. He was no doubt attracted by golden specks of the metal to be found in river gravels where he searched for his pebbles. We have evidence that Neolithic man used these gold specks and hammered them into ornaments.

Gold is a remarkable metal and is justly called 'noble' as it will yield easily to pressure and yet resist oxidation even in thin films. Consequently, when mineral veins carrying specks of gold are disintegrated by erosion the gold survives and becomes concentrated in river gravels and sands, and along the sea shore. Although thinly distributed in primary rocks this precious metal can be found throughout the earth's crust and most countries possess it to a greater or lesser degree. There is even gold in sea water and several attempts have been made this century to recover it. At the end of the First World War, Dr Fritz Huker, a German chemist tried in vain to fulfil his claim that the North Sea would yield sufficient gold to repay the war debt. Since the North Sea in places contains 5–10 mg of gold per ton of sea water this idea will no doubt be revived in the future.

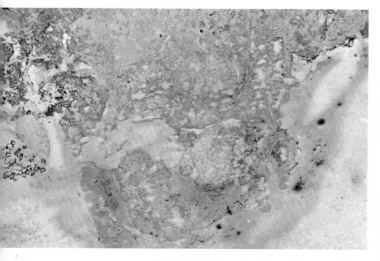

'Fool's Gold.' A vein of iron pyrites running through blue copper ore.

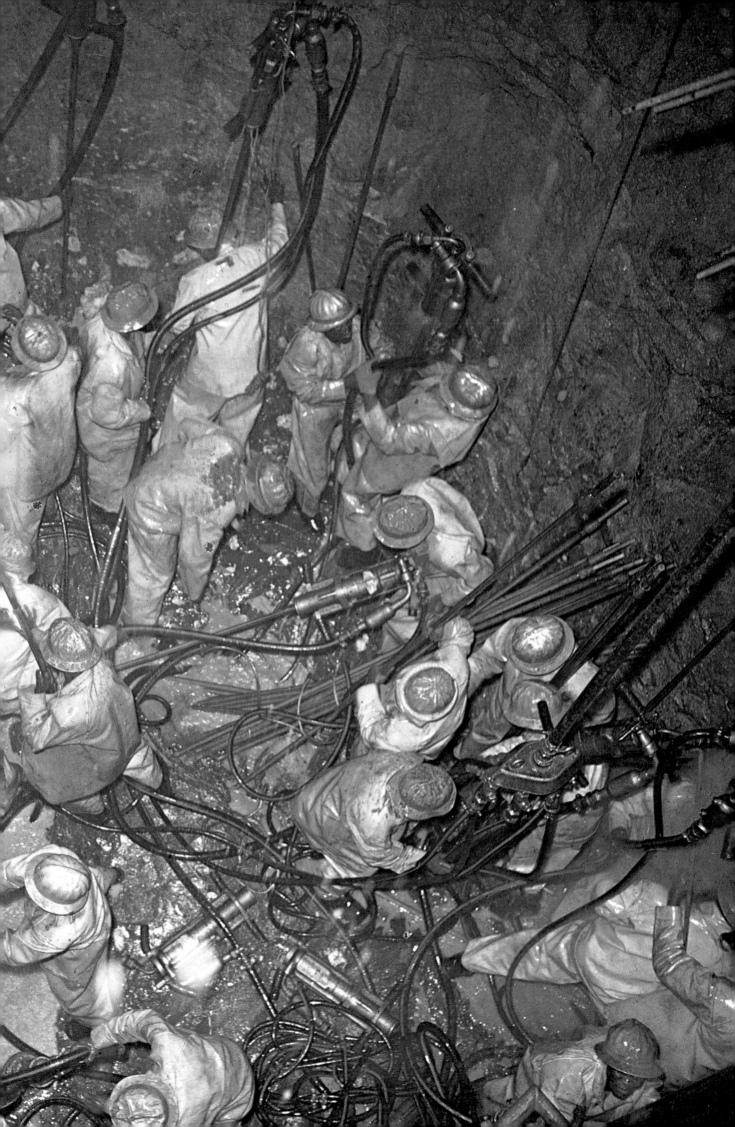

A typical rig used for mineral exploration. The geologist is holding a cylindrical core of rock withdrawn from the adjacent diamond-tipped drill pipe.

Just like the 'old timers' the ancients devised all manner of ingenious methods to wash gold out of sands and gravels. It is conceivable that the fabulous Golden Fleece was derived from the use of woollen materials for this purpose since gold will cling to matted fabrics. As a decorative metal it immediately acquired a significant place in the material order of society. Nearly three thousand years ago Croesus, King of Lydia, struck gold coins and since then the currency of the world has been based on gold. In 1821, Britain declared that in future all her currency would be backed by gold and within fifty years almost every other country in the world followed her example in establishing a gold standard.

The great strike of gold-bearing rocks in California in 1845 was followed three years later by a similar discovery at Bendigo in Australia. This gold rush was followed by the mad exodus of thousands of people into the Klondike river area of the Yukon. At a place called Bonanza in 1896 gold was discovered in the river gravels. Countless men and women disappeared in search of a 'bonanza'. Some survived and a few became rich; others were probably devoured by bears as they heedlessly excavated the quartz veins – most of which were barren of gold. But the greatest goldfield of them all had yet to be discovered.

Gold-bearing veins were known to occur in Africa and a seemingly rich strike was made at a place called Barberton. But these were difficult to find and hard to mine. It was not until 1886 that other types of gold-bearing deposits in South Africa were discovered. Over the course of millions of years sediments eroded from the high tablelands were poured by rivers into an enormous basin. The effect was to weather gold out of the veins and transport it along with sand and gravel into this basin. The gold became concentrated along an extensive coastline and then buried beneath thousands of feet of sediment supplemented by volcanic lava. Subsequent elevation of this basin and the eventual erosion of the land surface led to the re-exposure of the gold-bearing deposits now converted into indurated beds of quartz and conglomerates. These are now known as the 'banket'. This in broad outline was the geological setting for the development of the Witwatersrand Basin which contained the most widespread deposits containing gold yet known in the world.

The Witwatersrand Basin was discovered when a prospector, George Harrison, found gold in a bed of conglomerate. He sold his claim in 1886 for £10 and vanished from the scene without realizing that he had opened the door to a new era in the history of South Africa and a new geological attitude to the occurrence of gold. It needed the intuitive ability of men like J. B. Robinson to realize that Harrison's gold was not contained in veins descending into the ground like those at Barberton or at Bendigo in Australia. He shrewdly noted that these were sedimentary deposits which produced poor soils and so he proceeded to purchase at low cost extensive areas of seemingly worthless farm lands over the velt. Robinson eventually gained possession of land which realized £500 million of gold.

Not far behind him came Cecil Rhodes. It is typical that hunters and explorers had crossed the Witwatersrand (The Ridge of White Waters) without realizing the significance of Harrison's find. The Rand, the abbreviated name for this area, eventually produced half the new gold of the world, from an area about 50 miles long by 20 miles wide. In its development fantastic obstacles were overcome by mining skill and endurance against great odds as the shafts descended over $1\frac{1}{2}$ miles into the earth to reach the banket.

Geological circumstances caused the miners to sink shafts rather than send in adits from the surface along the gold-bearing beds. Excavating on this scale had never been done before. The hand drills used for holing the rocks to receive explosive charges demanded human labour on a large scale. The dust created entered the driller's lungs and hundreds died from silicosis. As the hole went down the unsupported rocks would suddenly burst, and the accident rate was high. But the lust for gold provided the incentive to press on regardless of the toll in human life. When the gold-bearing banket was reached huge horizontal tunnels were driven under similar hazardous conditions. Yet for some inexplicable reason men were prepared to descend into these death traps to extract this purely artificial symbol of wealth for others to live in safety and in luxury. Time and again one finds this characteristic of mining and miners.

In 1932 two young men – Emmanuel Jacobson, a solicitor, and Allan Roberts, a dental mechanic – gave up their jobs to realize the products of their hobby – their passion for minerals and mining. Roberts was a born prospector. A chance meeting with a prospector, Archibald Megson set him off in search of gold. Megson had sunk a shaft, 100 feet deep, single-handed into deposits which he thought were identical with those of Witwatersrand – on a farm called Aandenk in the Orange Free State. The rocks showed good values but he could find no one who was prepared to finance a venture so remote from the Rand. It was not the shaft which interested Roberts. He traversed the unpromising table-land and found conglomerates which convinced him that this was a geological extension of the Rand. He and Jacobson sunk every penny they owned or could borrow into a borehole at Aandenk. The drill bored its way through the conglomerates into hard green crystalline rocks, now known as the Ventersdorp lavas. At a depth of 2,721 feet the drill pierced the lavas and descended into volcanic sediments.

Contrary to all advice Roberts and Jacobson pressed on but money was running out. At 3,844 feet they had promising traces of gold. Down continued the drill to 4,064 feet but at that point the money ran out and work ceased. Other prospectors entered the region and sank equally fruitless boreholes, but no one had the insight to deepen the Aandenk Borehole. War broke out and prospecting practically ceased. Then eleven years later the borehole was deepened and within 400 feet it hit the gold-bearing Basal Reef of the Rand. Allan Roberts

Diamond hunting along the African coast. When raised beach deposits have been broken down diamonds are found lodged in the crevices and among the boulders of the ancient shore line.

died penniless and his friends subscribed to pay for his funeral. For another small expenditure he would have found what became the Loraine Mine. This is but one of thousands of anecdotes which exemplify the difficulties confronting even expert geologists when exploring a sedimentary basin as large and as complicated as the Rand.

Copper Mining

Fascinating though the story of gold might be, it was less noble metals like copper, tin, lead, zinc and iron which directed the development of civilizations. Copper was of much more practical importance to early man than gold. Like gold, copper occurs native and was found in disintegrated rocks in volcanic areas and was in widespread use in many parts of the world by the year 3000 BC, especially in the Middle East. Early craftsmen discovered that the native metal could be beaten with stones into sickles and daggers which were capable of cutting corn or flesh because the metal became quite hard when hammered. About the 3rd millennium BC, it was discovered that copper could be released from certain bright blue stones in a charcoal fire. It was probably the bright blue and green colours of copper ores which first attracted attention and the lighting of fires against such mineral deposits produced the molten copper. As a result, the technique of 'fire setting' became a standard method of splitting rocks. The modern use of flame throwers in quarrying is a product of this ancient technique of mining and smelting.

As time went by the surface exposure of copper ores were soon consumed and man was forced into the depths where the ore deposits became harder and more difficult to smelt. Copper ores, because of their vivid colours, were easy to recognize and early man would have had no difficulty in following them underground. The mines of Metterburg in the Tyrol were apparently worked from 1600 BC to about 800 BC. Likewise the copper mines of Cyprus were in full use in 1500 BC when they were supplying metal to Egypt. In all these ancient mines the ores were carried to the surface and pounded with stone hammers and then panned in two-handled pans. The concentrates were then put into clay-walled smelters. Inside the powdered ore was interbedded with layers of wood fuel which reduced the metal and caused it to run out into clay moulds. This illustrates how early in history man discovered the basic principles of furnace techniques and of wood-charcoal as a metallurgical fuel.

Bronze

Since ore-bodies are never composed of single minerals the product of these mines contain metals other than copper. When smelted some copper emerged as an alloy. Eventually it was found that when tin was involved the emergent alloy was harder and more easily prefabricated than copper. This was bronze – a metal which could be poured into casts and the resultant knives and utensils were much harder than copper. The alloy characterized a new era – the Bronze Age.

The manufacture of bronze became widespread not only in the Western World but also in the Far East. There are some remarkable accounts of the production of bronze during the Chin dynasty (1122–255 BC). Indeed so great was the respect for ancient bronzes in China that when an old ming (a tripod cooking pot) was found in the banks of a river in Shensi in the year 116 BC the name of the reign was changed in honour of this event. It is also said that the Chinese of 500 BC were further advanced in the use of metallurgy than the average European 1,500 years later. Be that as it may, the realization in all progressive communities of the value of alloys such as bronze gave great impetus to the mining of ores. All kinds of rocks were tested by smelting, and without any knowledge of the chemical properties involved it is truly incredible how few metals escaped discovery. Silver, lead, zinc and even mercury were commonly used and so it becomes impossible to state with exactitude at what point in history each was developed.

Drilling blast-holes in the copper deposits of the U.S.S.R.

The Search for Precious Stones

Again and again we see the primitive acquisitive instincts of man expressed in what can only be described as a lust for material possessions. Gemstones, and especially the diamond, in the early days had no conceivable practical use. The rarity and beauty of the diamond has afforded it a unique and exotic position in history. All natural gems and precious stones, with the exception of pearls, corals and amber, are inorganic minerals. The others are made by organisms. These gems are usually single crystals which are bounded by plane surfaces in a definite regular plan which we now know is an outward expression of the internal arrangement of the constituent atoms. There are six crystal groups or geometric systems to which crystalline minerals belong. The diamond belongs to the cubic system and it usually takes the form of an octahedron or a dodecahedron which makes it not only easy to identify but also endows it with natural breakage-planes called cleavages which facilitate cutting it into 'diamante' form.

In general gemstones originate in three ways. Opal, agate and some quartz crystals are formed by the deposition of gelatinous silica from aqueous solution in cracks and cavities in rocks and especially volcanic rocks. Diamond, ruby, sapphire, topaz, tourmaline and others

Digging for diamonds. The depth of the Great Hole of Kimberley forced individuals to combine in their hunt for diamonds in the blue-ground.

crystallize out from molten rock often at great depths in the crust. Thirdly, emerald, garnet, spinel and such gems develop in rocks through the action of mineralizing gases ascending into them from below. Since gemstones must be hard as well as beautiful, they tend to stand out on weathered rock surfaces. They also resist erosion and are very commonly found second-hand as pebbles in river gravels or on the sea shore. Beautiful though gems are in their natural form, their value is greatly enhanced by cutting. The oldest form and fashion is the rounded so-called 'cabochon' and this is still used to produce gems like 'tiger's eye' or to enhance the colour of opal. In this way too the pearly lustre of moonstone is achieved or the asterism exhibited by star rubies and sapphires.

Cutting gems into geometric shapes demands great skill and knowledge of the gemstone's crystal structure. In the case of the diamond it is possible to split it with a sharp tap on a chisel, provided the edge coincides with one or other of its eight cleavage planes. This will produce an octahedral shape. To enhance its light-reflecting powers the corners must be sawn off and then each set of corners so produced removed likewise. Since diamond is the hardest material know this can only be achieved by using another sharp-edged diamond or by abrasion with diamond powder.

The pioneer in the art of diamond cutting was a seventeenth-century Italian called Peluzzi who was the first to produce the 'brilliant cut'. In 1914 Marcel Tolkowsky put the art of Peluzzi on a more scientific basis by working out the precise geometry of removing corners from the double-sided pyramid of diamond. As a result the light entering the diamond is broken up into the rays of the spectrum and reflected out again by the facets on the underside, to give it its 'fire'. So, to recognize a well-cut stone one should examine its underside which should be as dark as possible with a pin-point of light in the middle.

There are many other diamond cuts and shapes. There is the Oral, the Pear, the Baguette, the Marquise, the Magna, the Royal, the French, the Full Dutch Rose, the Cairo Star and the Brabant Rose.

Diamond is simply crystalline carbon, but to produce it Nature has marshalled the full range of physical forces at her command in the depths of the earth. Artificial diamonds are produced if carbon is subjected to temperatures ranging up to 2500°C under pressure of between a half a million and one and a half million pounds to a square inch. What then must the conditions be like in the earth to produce a crystal as large as the Star of Sierre Leone (969 carats), the Excelsior (995·2 carats) and the largest of them all, the Cullinan (3,106 carats)? The process clearly requires immense heat, a confined space in which to create the pressures and a source of carbon. Heat and pressure of this magnitude are a commonplace occurrence in the crust, but these physical conditions are seldom associated with the right type of carbon – hence the rarity of the diamond. Given these conditions the crystallization is best achieved through the develop-

ment of inverted conelike structures known as 'pipes' many miles deep into the mantle of the earth. The most famous of these is the Big Hole of Kimberley. As far as we know only fourteen countries outside the U.S.S.R. have developed diamond pipes and of these South Africa, Sierre Leone and Venezuela have yielded the greater part of the world's output of natural diamond.

Being the hardest of known minerals diamond resists erosion and is therefore more commonly found in gravels eroded from pipe formations and transported great distances from their source. It is estimated that over 3,000 feet has been eroded from the Kimberley area to reveal the 'blue-ground' in which the diamonds were found and a great deal of it was conveyed to the sea by the Orange River. This is why diamonds are found at the mouth of this river and along the coast – a discovery which illustrates the way in which the minds of geologists work in observing simple phenomena.

The late Dr Hans Merensky, who had already become a millionaire through his discovery of the 'platinum reef' on the Transvaal Bushveld, heard that diamonds had been found on the treacherous coast of West Africa. He flew out there and observed that the diamonds had been discovered in beach deposits about ten to twenty feet above sea level. This was what geologists called a 'raised beach'. Such earth movements are rarely localized and so he was able to follow these beach deposits southwards along the Skeleton Coast of the Atlantic to the mouth of the Orange River. This he concluded was a potential diamond field containing diamonds washed by longshore currents along with the sediments draining into the Atlantic from the hinterland. His deduction was correct and he became a multi-millionaire as the owner of the largest diamond field in the world bounded on one side by the sea and protected on land by impenetrable desert.

In post-war years diamonds have been used more and more in industry. They provide the only known natural material for the cutting and fashioning of hard metals. So vital has it become that geologists as well as economists are able to define the level of industrial development in terms of diamonds consumed. This accounts for the vast sums of money which have been spent by industry on research into methods of producing artificial diamonds. To illustrate the progress which has been made to emulate nature, the U.S.A. used to import 12 to 13 million carats of diamond. By 1968 she had reduced imports to about $6\frac{1}{2}$ million carats by producing about 11 million carats of artificial diamonds to cover her industrial needs.

Iron

If diamonds are essential to modern industry so too is iron just as it has been since the Industrial Revolution. Fortunately, for every ton of rock in the known crust there is on average a hundred weight of iron. Next to silicon and aluminium iron is the most abundant of all elements. Even so, it is not easy to extract from all iron-bearing ore

The massive removal of the raised beach
deposits of South West Africa uncover
diamonds on a greater scale than anywhere
in the world.

The Star of Sierra Leone, one of the world's most famous diamonds.
Diamonds all. (middle) The wide variety of colours are created by the impurities in the stones.
(below) Crustals of opal: the iridescent colours are created by the reflection of light through layers of silica contaminated by organic matter.

deposits. Iron in its pure form lacks the hardness of bronze and only when it was discovered that by adding such substances as carbon or manganese to toughen it did iron supersede bronze as the everyday manufacturing metal. Only 500 years ago were furnaces created to make cast iron and until engines and railways were built there was no great need of it. Up until the eighteenth century the mineral wealth of the world, which was still based on agricultural developments, was dominated by the use of metals for utensils and decorative purposes. So gemstones like rubies, emeralds and diamonds were of greater interest to man than cast iron. The manufacture of swords and armour dates back a long way so that steel, which was iron case-hardened in various accidental ways, was a metal of greater interest to the war-mongering nations than to those content with an agricultural way of life.

For two centuries now the core of the pattern of industrial development has been the production of iron from mineral deposits. These are widespread, but without fuel the lavish distribution of iron ores is useless. A special fuel is needed to produce iron from ores; this is coke which is created by roasting certain types of coal.

The British invention of coke from coal by Abraham Darby in about 1730 started the rapid development of the use of iron. Darby, a Shropshire Quaker, found that certain coals when heated created a hard sponge-like mass of carbon which produced better smelting conditions than charcoal or coal itself. It was a timely discovery for in 1768 James Watt invented the steam engine. Between them these two men could claim justifiably to have initiated the industrial age.

Iron ores occur either as beds of sedimentary rock or as irregular shaped ore-bodies composed of the iron oxide mineral called haematite. Being richer in iron the ore-bodies became the main centres of mining activity. From surface exposures these mines eventually were forced underground and large numbers of miners were employed to achieve the increasing tonnages needed to meet the voracious demands of the foundries.

In the course of time enormous underground workings were developed as the ore was often in massive form enclosed in firm rocks which needed little support. The English miners of Cumberland and North Lancashire produced large tonnages from bodies of haematite which had been formed over 200 million years ago by mineralizing solutions in cavities created in Carboniferous limestone. Today the workings look like huge limestone caverns. At Krivoi Rog in the Ukraine there is an enormous ore deposit of haematite which seems to have been developed as a ferrogenous mantle over an old land surface. From this deposit outputs of over 20 million tons per annum are achieved, and the same kind of productivity has been reached from the Jurassic iron ores of Alsace-Lorraine – a deposit of immense political importance to the militant powers of Europe who have over the ages raided these mines in pursuit of self-aggrandizing wars of conquest.

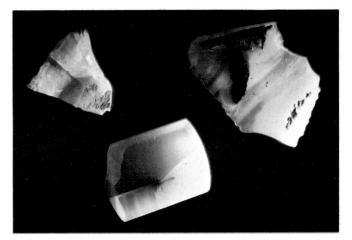

Wooden posts support the treacherous roof as miners remove the coal.

Steel supports enable machines to excavate the coal.

A mechanical excavator removing the overburden from seams of coal in the U.S.S.R. **(right)**

Coal Mining

Underground coal mining is the most hazardous of operations in the depths of the earth. Gold or nickel or copper mines may be deeper than coal mines but at depths of over one mile the dangerous pressures contained in the surrounding rocks are common to them all. But in the case of coal, all these mining hazards are supplemented by the release of poisonous, explosive gases and also dusts which petrify the miners' lungs. Gases compressed in the layered deposits which make up a coal seam are the remains of the volatiles created by the decayed vegetable matter. Undisturbed they are harmless, but once the surrounding pressures are relieved by the development of a coal face they become explosive. An accidental spark from a miner's pick, or the collision of particles of dust, can cause these gases to explode in badly ventilated sections of a mine. It is almost impossible to avoid such occurrences and no coalfield in the world has escaped the tragic effects of these explosive gases which are as eruptive as a volcano and by blast alone hundreds of miners have been killed. If they survive they are frequently trapped in the gas filled workings and die slowly from carbon monoxide fumes. The feats of bravery of miners involved in explosions rival those in wartime. Among this breed of men is an all-powerful element of comradeship which makes miners a race apart wherever they delve into the depths of the earth.

Opencast Mining

In addition to these massive accumulations of iron by mineralizing solutions, there are widespread deposits of sediment which contain enough iron to be classified as ores. Since these often occur as fairly flat-lying beds near the surface, they lend themselves to what is called opencast mining. Deposits of this kind are extensive and with the birth of engines it was natural to use machines instead of men to dig ironstone. The invention of the crane led to the idea that a scoop, or bucket, at the end of a levered arm would perform the same action as a man with a shovel. This gave birth to the mechanical navvy or digger. At first these diggers ran on rails which limited the positions in which they could excavate. Before World War I the caterpillar traction was invented and then used for the first tank designed to break through the German lines. The caterpillar track revolutionized mechanical excavators. Diggers so mounted could move freely within the opencast pit and discharge their loads into waiting trucks. They were, however, limited in their action since they could only dig above

34

A bucket excavator used in the construction of the Panama Canal, 1881.

A power shovel used for digging iron ore in Northamptonshire, England, 1881.

the level upon which they stood. As the bed of ore descended deeper into the ground the overlying rocks, or overburden, increased in thickness beyond the digging capacity of the excavator's levered arm. The concept of the crane reasserted itself and an excavator was made which had a bucket on a rope. By this means the excavating bucket was dropped from ground level, down the face of overburden and dragged upwards. The problem of the overburden was solved by the invention of the most versatile of all diggers – the dragline excavator. Not only was the face scraped away, but the broken rock filling the bucket could be flung outwards across the pit and discharged clear of the ore-bed exposed on the floor of the pit.

Black Diamond

Not all the coals around the world are suitable for metallurgical purposes. The term coal covers deposits which are the product of decay of forest debris buried over millions of years layer by layer in swamp lands. Obviously such vegetable deposits will vary in time and place. After burial beneath successive accumulations of sand and mud the changes wrought over millions of years have been exceedingly complicated and the end products equally enigmatic. Although the most erudite scientists of this century have studied coals they have failed to unravel all the secrets embodied in what is graphically called 'black diamond'. However, it is well established that only certain types of coal when roasted will yield coke.

Coke is now becoming increasingly difficult to produce and as supplies diminish at an alarming rate new fuels are being exploited to meet the demands for more and more metals, especially iron and steel. Even so, coke is still the most successful solid fuel for smelting metallic ores. By discovering coke, and the method of mass producing iron and steel, Britain led the world in its dependence upon coal. The British Isles still possesses the greatest resources of coking coal in the western world.

Black Gold

Oil was first struck at depth in Pennsylvania by Colonel Drake in 1859. In Scotland James Young had shown, somewhat earlier, that oily substances could be obtained by distilling cannel coals and shales. From these two events sprang the mighty petroleum industry. Initially oil was used for illumination – to replace colza oil which was then the principal lamp fuel. Oils were also used for lubrication so that the discoveries of James and Drake were highly opportune for the development of machines.

For a long time geologists depended upon surface seepages of oils to locate supplies. By experience, and the geological evidence produced by boreholes and fieldwork, it was found that petroleum was usually associated with deposits originating in the sea. From this it was concluded that natural oils were produced by the decay of marine organisms in contrast to the decay of plants which yielded

Explosion on Rig 20 in the Persian oilfields. When a well explodes and catches fire, the steel lining tubes are shot out into the air like coiling ropes.

Drilling for oil in an anticline. By striking gas the resultant pressure creates a gusher.

cannel and coal. Consequently, exploration was directed into lands which contained great thicknesses of fossiliferous shales and lime-stones. These were regarded as the source rocks of the petroleum. As such, the oils were evenly distributed through the sediments and in this condition would not flow into an oil well.

To achieve a flow another geological process was required to con-centrate the disseminated globules of oil into porous rocks out of which the oil could be pumped. It was found that when these source rocks were arched upwards by earth movements into folds called 'anticlines' the petroleum had been squeezed towards the crest. If porous sandstones were present in the sedimentary succession, these accepted the petroleum and concentrated it into reservoirs. To prevent the oil from escaping the sandstones had to be overlain with im-pervious rocks such as shales. The target for the oil geologist was a succession of source-rocks, reservoir-rocks and 'cap'-rocks folded into an anticline. There are several other ways in which oil can be accumulated but this structural situation has been found to achieve the best conditions for the development of large accumulations of petroleum.

A graphic way to illustrate the mode of formation of an anti-clinal development of natural gas and oil has been achieved by the construction of an oil model based on the geological evidence obtained from the discovery of the oilfields of Nottinghamshire in England. Without the knowledge of the way in which oil and gas had accumulated in Nottinghamshire it is doubtful whether this dramatic development of one of the world's largest natural gaso-meters would have materialized. Moreover, the geophysical methods of locating these anticlines beneath the bed of the sea would not have been available if similar oilfields on land had not been the product of over 50 years of research into the use of gravity and electrical currents and sound waves in probing the rocks to great depths beneath the sea.

The same structural situation exists in the Middle East where huge anticlines have been formed by the earth pressures which created the Alpine Chain on one side and the Himalayas on the other. Into these have seeped the petroleum which had accumulated throughout millions of years of sedimentation in an ancient sea called Tethys. It is doubtful whether another similar basin of sediments will again be found on land. The discovery of large oilfields in the future will lie in the shelf areas of the continents. Since these are composed of marine sediments it only requires earth pressures to squeeze and con-

A ladle of molten steel.

The unit construction of an oil tanker in Japan.

everywhere as the continental shelves have been, and still are involved in the huge plate-like movements of the crust created by internal forces which force up ridge-like wedges midway along the ocean floor. Thus the exploration of that hidden world by huge floating drilling rigs will be yet another contribution which the drillers for oil will make to the future development of the earth's resources of minerals and energy.

Politics and Resources

National frontiers rarely coincide with geological boundaries and the distribution of mineral resources has always been the incentive behind the expansion of a nation into an empire. The Egyptians did it; the Greeks did it and the Romans copied these patterns in a remarkable

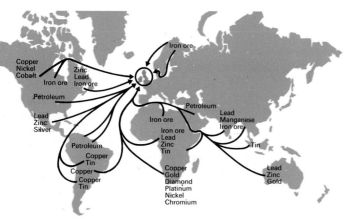

British World Wide Mineral Sources in 1939

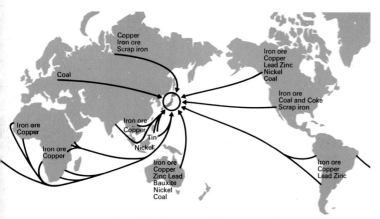

Japanese World Wide Mineral Sources in 1973

The trade routes of Japan compared with those of the British Empire in 1939.

centrate the oil into anticlines. This is likely to have happened almost way. However, it was not until the industrial revolution ushered in the machine age less than two centuries ago that minerals began to be used in large volumes. Both the machine and the power to run it required minerals and fuels. No nation has ever been fully endowed with these and the growth of the British Empire has become the outstanding example of the growth of power through the acquisition and exploitation of mineral resources.

Between the two world wars the growth of technology highlighted critical needs and it is now clear that certain minerals can be classified as of strategic importance. High on this list are the minerals which yield manganese. Without it the production of steel would practically cease. Manganese removes oxygen and sulphur from molten iron as it passes into steel. The U.S.S.R. is the world's greatest producer, next comes India, then China. This illustrates the picture of all the other so-called strategic minerals. No single country has a full quota of minerals to make it totally self-sufficient. So much so, if the stockpiling of minerals by Germany, Italy and Japan had been assessed correctly the prospect of war in 1939 would have been anticipated much earlier.

Ironically, much of the strategic minerals were supplied to these militant countries by Britain and France. The U.S.A. also entered the picture by supplying substantial quantities of copper to Germany, Italy and Japan between 1935 and 1939. She also supplied Japan with nearly 9 million tons of iron and steel and 100 million barrels of petroleum products. From this it should be clear that the stockpiling of strategic minerals and metals by one power is a clearer indication of military build-up than the political charades of national leaders and diplomats.

This concept was realized at the dawn of this century and from it emerged the study of geopolitics by the German, Friedrich Ratzel, and the Swede, Rudolf Kgellan. According to these writers, whoever controlled the great land masses of Eurasia and Africa would have at their disposal enough material resources to control the seas and thus the world. These land masses were called the 'heartland': it was judged that the best centre of control would be in Western Europe. Hitler borrowed this idea and used it as an operational concept in 'Mein Kampf'. Today the picture is much more confused, as the importance of certain restricted sources of strategic minerals, especially radioactive ones, endows many small countries with strategic control over supplies.

Burning lava spills out over the flanks of Mt. Etna.

Chapter Three

THIS VIOLENT PLANET

A Moving Crust

To have ever seen an earthquake or witnessed a volcano erupting will give some idea of just how great is the energy contained in the bowels of the earth. Both these phenomena are uncontrollable yet without them this planet would probably have exploded long ago into myriads of meteorites. Earthquakes and volcanoes act as safety valves for the built up pressures within the earth's interior. In less dramatic ways the energy of the earth is also dissipated by infinitesimally slow movements and adjustments to the crust – a continuous process operating on the solid crust from inside. There are now instruments which can measure this process and they have revealed some startling facts: the materials beneath the crust are not always red hot or even liquid. Measurements have shown that there are both hot-spots and cold-spots and that these coincide with materials which behave as liquids or solids. For want of a better term these sub-crustal substances are known collectively as 'magma'.

Magma becomes molten when the pressure exerted on it by the crust is relieved by earth movements. Otherwise these magmatic materials seem to behave as hot solids within the recesses of the earth. When magma forces its way upwards it creates some of Nature's most beautiful artefacts. For example, when it invades a mass of limestone it is converted into variously coloured marbles. If the limestone is pure calcium carbonate white marbles are formed like those of Carrara in Italy. When limestone containing impurities is invaded by magma various colours are induced like the serpentine-marble of Connemara, Ireland.

Much of our knowledge of magma comes from information about vibrations recorded by seismographs during earthquakes.

Catastrophes from History

In describing earthquakes and volcanoes there is little danger of indulging in overstatement as they are the most dramatic events on earth. Possibly the greatest recorded earthquake disaster was the one which occurred in the sixteenth century on the Hoang Ho Plain of China in which 800,000 people are said to have lost their lives.

A typical pattern of an earthquake can be seen in the catastrophe at Lisbon on All Saints Day, 1875. The churches were thronged with worshippers when suddenly there was a terrific roar and the roof and arches of great stone churches crumbled in the grip of jarring shocks. Within ten minutes nearly a quarter of the population of Lisbon had perished and thousands more lay trapped in the ruins. Records show that the sea withdrew from Lisbon harbour and then rushed back as a tidal wave fifty feet high swamping hundreds of survivors who had taken refuge on the wharves. A shorter quake followed raising enormous clouds of dust out of the debris. Two hours later a third shock wracked the stricken city. The account goes on to show that the vibrations were not confined to Lisbon alone – another characteristic of tremors. Records show that huge waves

Earthquake damage in Caracas. Nothing can withstand the disruptive forces of an earthquake.

Earthquake in Peru. A yawning gap in the road leading to the stricken village of Casma.

a The town of Yungay, Peru, dominated by the country's highest mountain, the snow and ice cornice of Nevados Huascaran. This picture was taken before the earthquake of 1970.
b Yungay after the earthquake. The peak of Huascaran has broken off. It fell 1,000 meters and started an avalanche which demolished the town.

The eruption of Mt. Pelée, Martinique, in 1902 which created an avalanche of lava, ash and gas.

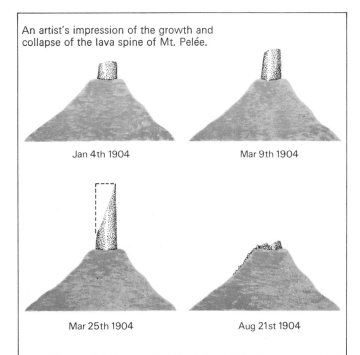

An artist's impression of the growth and collapse of the lava spine of Mt. Pelée.

Jan 4th 1904

Mar 9th 1904

Mar 25th 1904

Aug 21st 1904

splashed across Loch Lomond in Scotland, 1,200 miles from Lisbon. The river at Lubeck in Germany rose several feet, 1,400 miles away from the earthquake centre; and in Fez and Martinique, Morocco, 400 miles away, the damage was almost as severe as at Lisbon.

Earthquakes are not confined to the warmer regions as the great earthquake which wrecked south-eastern Alaska shows. A long series of shocks heralded the devastation centred on the sparsely populated region of Yakutat Bay. Some of the greatest displacements of rocks and ice yet recorded were created by this quake. The coastline was lifted in places more than 47 feet and huge icebergs split off from the snout of the Muir Glacier. Inland the Malaspina Glacier, which had remained inactive for so long, enabling a dense forest to grow upon its surface, was shattered by this shock and the ice came to life. The surface of the glacier was tossed into wild confusion creating a day-of-judgement picture of overturned trees tumbling into deep crevasses and the erection of mountains of ice blocks.

As well as lifting coastlines earthquakes can sink them, as happened on December 28th, 1908 when eastern Sicily was struck by a quake which destroyed the cities of Messina and Reggio, killing over 100,000 people. When the shocks ceased it was found that the coastline had submerged two feet. Accounts such as these have become classic in the folklore of natural disaster, but they ignore those which have constantly taken place in areas where no substantial loss of life has occurred but where the geological effects have been just as vivid.

Because of its fame as a city the earthquake which struck San Francisco a little after 5 p.m. on April 19th, 1906 has been documented and studied in immense detail. As in Wellington, New Zealand, the inhabitants of San Francisco believe, and probably rightly so, that another quake will come. About 700 people were killed – miraculously few under the circumstances – and immensely valuable properties crumbled or were destroyed by the fires created by burst gas pipes and severed power lines. The broken water mains left the firemen helpless and the city was reduced to a complete shambles. The earthquake was triggered off by movement along a great fault zone known as the San Andreas Rift. It extends obliquely along the Californian coastline for over 600 miles. As a result of the quake the ground was ripped open along these faults for more than 270 miles as the land on either side heaved sideways. In places these lateral displacements measured over 20 feet – dramatically demonstrated by roads which were cut and offset. The worrying feature of this type of earthquake belt is the length of time it takes for the region to attain stability. Minor shocks occur constantly and sometimes these culminate in a severe jolt such as the one which hit Imperial Valley in December 1940. In the orange groves the lines of trees as well as the roads were displaced sideways as much as 15 feet as the rocks slipped southwards on one of the faults of the San Andreas Rift as it plunges into the Gulf of California.

Submarine Quakes

Contrary to popular belief earthquakes do not only occur on land. Quakes under the sea can have horrifying consequences for the nearest landmass. Waves caused by submarine earthquakes are not often recorded but sometimes they reach dramatic levels such as the 93 feet high wave which struck the city of Miyako, Japan in 1896, or the wave which carried the U.S.S. *Watersee* far inland from its anchorage off the Chilean coast in 1868. Usually such waves are caused by sudden subsidence of the sea floor which withdraws huge volumes of water in a few minutes and then expels them as gigantic fast-riding waves. These are by no means localized events. Waves created by an earthquake in Peru were still eight feet high when they reached Japan 10,000 miles from their source. They must have been travelling at over 400 miles per hour – often the speed at which such waves move as a result of a submarine earthquake. In the open ocean their effects are not noticeable as the speed of the waves is influenced by the depth of water. The damage they cause to islands like Hawaii has been so great that coastal warnings are now broadcast when tremors are recorded by seismographs thousands of miles away.

Originating beneath the sea the displacements caused by these tremors are not usually detectable, but when they affect a coastline and its continental shelf we know what effects they will have. One such set of circumstances has provided a clue to the geological origin of deep valleys, called submarine canyons. Most of these canyons are situated on the continental shelf but the gorge often begins far from the shore. Being well away from the mouths of the rivers they are obviously not extensions of these terrestrial arms of erosion. The mystery of their origin was cleared up when it was discovered that an earthquake in the Grand Banks of Nova Scotia created submarine rivers of sand and mud possessing enormous erosive powers. The force of the quake cut the transatlantic telegraph cables which were involved in this turbidity current. Thirteen of them were severed during the thirteen hour duration of the earthquakes. From these accurate records it is calculated that the river of sediment travelled at the rate of 58 miles per hour for a distance of 295 miles.

Like glaciers these sedimentary rivers erode straight canyon-like valleys into the submarine shelf. Some, like those off the coast of California, are 150 miles long and 3,000 to 5,000 feet deep. It is now believed that a large number of these vivid submarine canyons are eroded by turbidity currents of sand and mud triggered off by sudden earth movements like earthquakes. There is no doubt that these tremors are as devastating below sea level as above it. The earthquake near Disenchantment Bay, Alaska in 1899 caused the sea floor to rise 47 feet above sea level and many others have destroyed coral reefs and created islands in many parts of the Pacific Ocean.

Earth tremors happen somewhere round the world every hour of the day in response to movements of the earth's interior. Their occurrence is so capricious that no one set of circumstances governs their behaviour. But one thing is common to them all. They always occur where the crust is either brittle or fractured by faults. And they always seem to be associated with areas which have either been activated by volcanic eruptions or are about to be.

Waves of Energy

From the selection of historic anecdotes quoted above certain features of the records reveal important characteristics of earthquakes. All the events take place at high speed. First air pressures and noise develop. Sometimes they express themselves like the roar of a passing train and at others as the whistle of a jet aircraft. This is due to the fastest travelling vibrations set up by rock fractures, namely sound. Sound waves audible to the human ear travel at 331·6 metres per second and are therefore the first indication of the impending disaster. In the ground the combined forces of these vibrations, called seismic waves, travel at differing speeds which can be measured by seismographs. The data they produced has revealed that each tremor broadly speaking contains waves of energy which vibrate in two main directions at right angles to each other, causing waves like those at sea. These waves travel out in all directions so that measurements taken by seismographs at different places provide the data to pinpoint the intensity, the direction and the origin of the quake. This technique has been proved by taking seismograph records of explosions detonated in the bottom of boreholes and underground mines. As a result of this kind of experimental work it is now possible to relate the speed at which these vibrations are transmitted to the nature of the rocks they penetrate, and so attain a great deal of information about the nature of the earth's interior.

When rock formations suddenly snap the release of energy is enormous; the vibrations which emanate from these ruptures in the solid crust usually resolve themselves into three broad types moving as rapid waves: (1) P-waves, speed waves in which the energy vibrates along the line of travel. (2) S-waves, vibrating at right angles to the line of travel. (3) L-waves, more slow-moving, which are concentrated in the near-surface areas of the crust. The fast P-waves exert a sort of push-and-pull action on the rocks, while the S-waves shake the rocks by the sideways movements of their vibrations. Situated on the near-surface rocks the L-waves tend to follow a zig-zag path. But all three waves are moving fast so that in many cases only the refined response of the seismograph can detect the individual contributions of each type of wave front to the disruption of the earth's surface.

These waves can travel immense, and in some cases, almost limitless distances. Although certain of the waves are completely damped out when they enter liquids. Consequently when it was found that they failed to penetrate the earth's core geologists concluded that the centre of the earth must be liquid.

These seismic studies also brought out the surprising fact that the solid crust was not floating on molten rock since the vibrations carried on down to the liquid core. From this it was deduced that the materials separating the core from the solid crust were more or less solid. This has since been referred to as the 'mantle'. So curiously the mantle, while being somewhat solid, can become liquid under certain conditions to yield the lavas erupted by volcanoes.

By making a close study of the rate of travel of these vibration waves a Yugoslav seismologist, Andriga Mohorovicic, found in 1909 that these waves changed speed at a certain level in the mantle. From this he concluded that there were certain points in the crust where the nature of the materials changed abruptly.

By joining these points Mohorovicic found that the division between fast and slow speeds created a kind of contour line inside

Aftermath of an Icelandic volcanic eruption. The rapid build-up of ash as the marine volcano of the Westerman Islands exploded in 1972.

the earth which has since been called the 'Mohorovicic discontinuity'. For convenience this term is now popularly known as the 'Moho'. By plotting the position of the Moho as the base of the crust it has been found that it varies in depth from 18 to 27 miles. Beneath continents and mountain chains the depth of the Moho is nearly twice as great as it is under the ocean floor: it must therefore represent a fundamental characteristic of the more mobile parts of the earth. This will be discussed again later, but as an insight of the earth's internal energy it must be a zone in which most of the gaseous, thermal and radioactive sources of energy must lie to trigger off earthquakes or create volcanic eruptions.

From recent experiences it might be claimed that the incidence of earthquakes has increased in the past half century. There is no real body of evidence to support this. On the other hand, their catastrophic results have become increasingly magnified by the growth of cities all over the world. Loss of life is certainly on the increase as people are herded into closely built housing schemes or high rising buildings. These are natural victims and are liable to be damaged even by minor earth shocks. Despite this the planning or maintenance of city life ignores the earthquake hazards of faults in the ground or the nearness of volcanic eruption. Movements which take place continuously along the celebrated San Andreas Fault have not inhibited the construction boom in San Francisco. The destruction of 200,000 people in the province of Kansu in China by an earthquake in 1920 did not deter people from going back to live there. Yet seven years later another earthquake occurred and killed a further 100,000 Chinese. Today the area is still heavily populated. Managua, the capital of Nicaragua, is apparently to be rebuilt despite being repeatedly savaged by earthquakes associated by vulcanicity. Some perverse element in human behaviour seems to compel people to remain as long as possible in one place regardless of the natural hazards which they know to exist there.

Greatest Show on Earth

Living on the flanks of a dormant volcano like Vesuvius or Etna has economic incentives which evidently outweigh the dangers of eruptions. The soil on the slopes of volcanoes is enriched by the ash and lavas and is excellent for growing vines and subtropical fruit. As with earthquakes, it is as yet impossible to predict the time of a volcanic eruption. Both events are the product of earth movement and that these are generated deep down by activity in the mantle. Like earthquakes the dramatic eruption of a volcano has been a favourite topic of doom-ridden historians.

In practically every eruption the breakthrough is centred about deep-seated cracks in the solid crust. The initial breakthrough is usually quite unspectacular as was shown in 1943. One day a Mexican farmer suddenly encountered a sudden spout of steam. This seemingly harmless event continued for a day or so when suddenly

Old Faithful, Yellowstone Park, U.S.A., which spouts steam to an hourly timetable.

the ground erupted and so was born the celebrated volcano of Para-cutin which now has a cone over 1,000 feet high. One day it will subside, but like all volcanoes it is liable to erupt time and time again.

Vesuvius had slumbered for centuries and vineyards clothed its fertile slopes. Suddenly in the year AD 79 after a short prelude of destructive earthquakes it exploded, blew off the top of its caldera burying the inhabitants of Pompeii in hot ashes and swamping the city of Herculaneum with hot muds and lava. This eruption, like others, was caused by compulsive movement along a line of weak-ness – in this case, the brittle axis of the Mediterranean. The history of Etna and Stromboli is the same; these are typical of volcanoes which erupt along the flanks of mountain chains like the Alps on one side and the Atlas Mountains on the other. Between these two earth revolutions the rocks are bound to be in a state of tension and these are 'ideal' conditions for volcanic activity – hence the history of eruptions in the Mediterranean.

There is evidence to show that some of these eruption centres – volcanic hearths as they are called – seem to be located at depths of as much as 60 to 95 miles. More normally they have been found to be at depths of 30 to 40 miles, as at Klyncherskaya Mountain, the largest volcano in Eurasia. By contrast a recent survey of the Japanese island of Oshima showed that its magnetic field was normal all over the island. Shortly after this survey had been completed the dormant Mihara volcano unexpectedly came to life. It was then found the magnetic field had changed and that the depth at which the altera-tion had taken place was only $1\frac{1}{4}$ miles below the earth's surface. This is a mysterious effect which has yet to be explained but it is similar to what takes place on the ocean floor as a result of volcanic eruptions.

From these few instances it is clear that there are no clearly defined conditions for predicting volcanic eruptions. On the other hand, the quality of the lavas they produce will determine the way in which they erupt and the type of volcano which will materialize. The behaviour of the lava when molten and its composition when solid is governed by the amount of silicic acid the magma contained. On this basis volcanic lavas fall into three broad categories: acid, intermediate and basic. Wherever the geologist goes, whether it be to the Pacific coast, the Ural Mountains of Russia, Central Africa or the Americas, he is able to tell by the type of activity whether the lavas are acid, intermediate or basic in composition. The acid lavas, being highly viscous, well up and move slowly so that the front of the flow is like a slowly advancing wall, bulldozing everything in its way.

Probably the most spectacular acid volcano recorded was that of Mount Pelee on the island of Martinique. It was an old volcano which had lain dormant for about fifty years. In the early spring of 1902 it came to life and the old crater was slowly filled with viscous lava. Eventually the subterranean forces beneath this heaving cauldron split the sides of the crater releasing magma and volcanic mud which descended at the speed of an avalanche.

A dramatic example of a mud volcano in the Buzan Valley of Rumania. The mud oozes out like chocolate.

The fissuring of the crust—an after effect of volcanic eruption.

St Pierre lay in the path of one of these *ardentes nuée*, as they have been aptly called, and within minutes the town of 30,000 inhabitants ceased to exist. A. Lacroix, a geologist, was there at the time and his verbal and photographic records have provided a remarkable set of facts concerning the eruption.

As the lava avalanches began to disappear he saw a plug of viscous lava being forced into the sky. Within seven months it had risen to a height of about 900 feet above the crater. Suddenly it split slant-wise down the middle leaving a thin spine towering into the sky like a church steeple. Unfortunately, this incredible monolith failed to survive; it crumbled to pieces within a year owing to the release of gases from the cooling mass of lava. Gradually the volcano became inactive and remained so for about twenty years, when the whole scene was repeated. Fortunately the re-built town of St Pierre escaped as the *ardente nuée* rushed northwards along valleys leading to the sea.

Volcanic activity can create new topographical features as happened at Krakatao in 1883. Up to that time Krakatao was an uninhabited dormant volcanic island in the Sunda Straits between Java and Sumatra. On the 26th August of that year detonations were heard at short intervals coming from the island and the sky was filled with dust and steam which completely obliterated the island. Next day four stupendous explosions were heard as far away as Australia – a distance of 3,000 miles. Eventually, when the island reappeared out of the dust clouds only a third of it remained. The rest had been blown into the sky and the dust travelled around the world several times. It was estimated that four cubic miles of rock were blasted into the air by the explosions. The tidal waves created swept 36,000 people off the low lying coastlines of Java and Sumatra. The volcano then remained dormant for 44 years before erupting again, but this time in a more normal way. For about six years this activity persisted and lava/ash cones were developed creating the new island aptly called Anak (child of) Krakatao.

Explosions of this kind are features of volcanoes which have lain

dormant for long periods of time. Mount Katmai, in Alaska, had slept throughout historic times, but in 1912 it exploded almost as violently as Krakatao. The steep-sided cone was shattered into an irregular mountain and the country around was buried beneath pumice, blocks of lava and tiny sharp particles of glass called 'shards'. On Kochak Island 60 miles away, forests were buried by

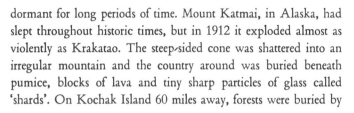

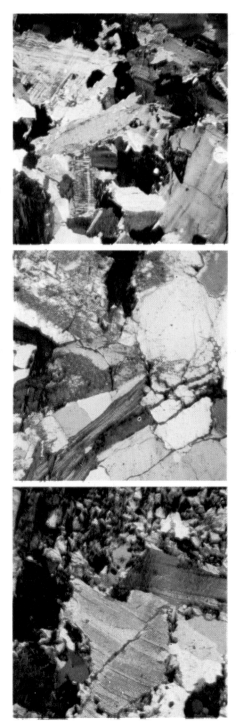

The granite tors of the Cheese Ring, Cornwall, England have been sampled and these samples have been ground until they were transparent. Under the polarizing microscope we see **(right)** three areas of the minerals in the granite.
Grey striped felspar crystals and dark and white quartz enclose the flakes of multi-coloured muscovite **(top)**.
Flakes of brown-green biotite mica with rounded areas of quartz and discoloured felspar **(middle)**.
The minerals have been broken as the granite finally cools into its ultimate solid condition **(bottom)**.

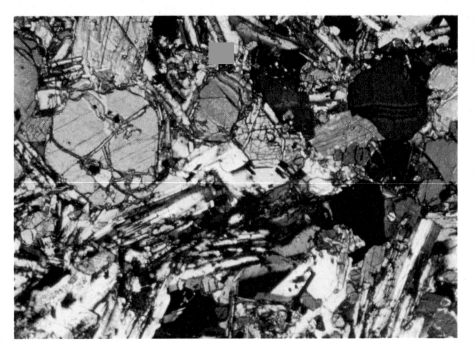

2a

Granites invaded by an intrusion of black dolerite have been displaced by a fault. The samples taken at the points indicated reveal 2a how crystals of felspar form as laths against the force of granite. 2b In the heart of the molten instrusion large yellow-brown crystals of olivine and agate have grown along with the crystals of the felspar. 2c The granite has been intruded by hot liquids carrying tin and tungsten minerals.

2b

2c

2b

this debris and even today the air is still laden with shards to such an extent that cars in the town of Anchorage wear out quickly owing to the intake of this abrasive dust into their engines.

The violence of these volcanoes which produce acid lavas and intermediate lavas suggest that they originate at great depths in the earth and are produced from magma which itself must have been rich in silicic acids. Moreover, it is clear that the nature of the crust above these silica-rich magmas may prevent such magmas from breaking through. This is particularly true of mountainous areas created by uplifting forces since they are found to contain cores of silica-rich magmas which have cooled down and crystallized into granite. Such cores of magma are revealed when mountains and rifted continental masses have been dissected by agents of erosion. A well-known example are the granite batholiths of Cornwall and Devon. These are the core of a mountain chain called the 'Cornubian Highlands' which was erected across Europe by earth pressures some 250 million years ago.

Likewise the crust is permeated by wall-like dykes of magma which form what are known as minor intrusives and usually occur as swarms filling fissures opened up in the crust but failing to reach the surface. These, like the granites, are of great economic importance as they often carry concentrations of valuable minerals such as quartz, feldspar and mica – especially the acid types called pegmatites. Among these minor intrusives are special ones called 'sills'. These are sheet-like intrusions which often extend through square miles of country rock. Fingal's Cave in Scotland and the Giant's Causeway in Northern Ireland are celebrated examples. Along the dikes they indicate the enormous amount of energy which is released in forming dike-like fissures and the separation of thousands of feet of solid rock to make way for their entry. This is accomplished by enormous volumes of super-heated steam and gases being released from the interior.

Perhaps one of the most vivid expressions of the earth's energy is given by the extrusion of basic magma to yield basaltic lava flows. Basalt is a dark-coloured, fine-grained rock. Rifts in the crust have released fantastic quantities of basaltic magma which, being low in silica-content, flows at high speed over vast areas. The Deccan Plateau of India was so formed and in Oregon the Columbia Plateau contains basaltic lava flows covering 50,000 square miles.

The Geography of Volcanoes

Two-thirds of currently active volcanoes, and immense numbers of dormant ones, are distributed around the borders and island festoons of the Pacific. In striking contrast the coasts of the Atlantic are comparatively free from volcanic activity. Where mountain-building movements, accompanied by earthquakes, are in progress no volcanoes occur as the elevating processes relieve the crust of its mounting

3a

The lavas of 'Ol Donyo L'Engai': the arrow
marks the spot where a sample was taken
for analysis. The microscope reveals the way
in which the molten rock has crystallized
with needle-like crystals by rapid cooling
3a and with large crystals inside the lava
slows **3b**. The ash in the background is
composed of the shattered crystals shown
in **3c**.

3b/3c

Above: Pillow lava on the mid-Atlantic Ridge 2,650 metres below sea level. This lava only forms when molten rock boils out on the sea floor.

Below: Pillow lavas as seen by a skin diver at a depth of 12 metres.

internal energy. Tibet seems to be free from volcanoes, presumably because it stands too high, but in less lofty areas of Asia, such as around the rim of the Tarim basin, active cones have been recorded. Similarly, volcanoes occur in Mongolia and Manchuria.

At the moment the only active cone in West Africa is Cameroon Mountain which forms part of an extinct volcanic chain which extended as a string of islands far into the Gulf of Guinea. However, from time to time the African continent has been pitted by numerous eruptions, some of them well preserved in the highlands of the Sahara. The scattered islands around Antarctica are also volcanic and on the fringe of that continent are the active cones of Erebus and Terror.

Where the crust is thin or where areas called 'plates' are moving apart the basic magma is liable to break through. This is particularly true on the ocean floors. These have been expanding for over 200 million years. In the northern hemisphere this has created eruptions which have built up Jan Mayen as an effigy of one of the world's greatest lava fields. It has developed on either side of a ridge extending from Britain to Greenland. Like Jan Mayen, the score of active volcanoes in Iceland are tapping the basaltic magmas upon which the plates are riding. The present eruptions of Helgafell Volcano in the Westman Islands will continue for a long time to come but no-one is yet able to predict how and when they will recur.

Mauna Loa

One of the most intensively studied volcanoes on the Hawaiian Islands is Mauna Loa. The last major eruption was in 1960, but it is by no means dormant. Mauna Loa rises 14,000 feet above sea level and at the summit the caldera floor lava is continuously on the boil. Twenty miles down the southeastern slopes of Mauna Loa is the sister volcano, Kilauea, with a caldera $1\frac{1}{2}$ miles in diameter. It too has a pool of boiling lava which in 1924 was for some unknown reason sucked back into the vent. Ground water poured in and caused a violent steam explosion. This was an actual demonstration of what seemingly must happen when magma encounters underground water. This is why volcanoes emit more steam than any other form of gas. These two craters standing nearly two miles apart, with Mauna Loa nearly 9,000 feet higher than Kilauea, appear to be fed from different vents or otherwise the lava pool of Mauna Loa would drain downwards to the lower crater.

These craters still contain magma so that the feeders which originally created the island are still open. Since the sea floor is 15,000 feet down it means that at least 29,000 feet of successive lava flows and ash have been formed in this one locality alone. This, in context with the yield of magma which has poured out to form similar islands, testifies to the existence of enormous earth movements created by the release of magmatic energy in the mantle beneath the Pacific Ocean.

Columnar jointed lavas poured out by
successive eruptions in Iceland.

Hot Springs and Geysers

Another expression of the earth's internal energy is the existence of
hot springs and geysers. When ground water descends to great depths
in deeply folded rocks it abstracts heat from the contorted deposits
and may re-emerge as hot springs. This is a widespread phenomenon
commonplace in areas which have been subjected to vulcanicity and
hot springs persist long after the volcanoes have ceased to erupt.
Sometimes the temperatures are great enough to convert the descending water into steam which is blown back into the sky as a geyser.
There are three volcanic regions in which geysers occur on an impos-
ing scale: Yellowstone Park, U.S.A., the North Island of New
Zealand and Iceland. The term geyser comes from 'gesir' the Icelandic name for the Great Geysir which is the most spectacular of
many steam spouts situated in a broad valley to the north west of the
Mecla volcano. These, and most of the other types of geysers, are
formed by the conversion of water into steam by subterranean hot
lavas causing it to build up pressure which spouts steam into the air.
As soon as the pressure is released the spout ceases, only to build
itself up again. This has been proved in Yellowstone Park where
'Old Faithful' erupts once an hour, sending a column of steam high

into the air. Borings put down in the park have encountered temperatures of 205°C at 245 feet which are more than capable of converting large volumes of descending ground water into steam. Since steam is a better solvent for rocks than cold water geysers create spectacular deposits and even form boiling mud-volcanoes. The Mammoth Hot Springs of Yellowstone Park are made even more spectacular by the large amounts of travertine which has been deposited by 'steaming' calcium carbonate out of the surrounding limestone. Mounds and terraces have been built up until eventually they look like petrified waterfalls. Sometimes the water is rendered alkaline by the rocks it passes through and when converted to steam it becomes capable of dissolving silica. This is deposited as siliceous sinter or geyserite – a form of opal. Basically white in colour, the geyserites develop the most wonderful range of tinted colours. Probably the most beautiful examples ever produced were partially destroyed at Rotimahama by the catastrophic eruption of Taraivera Volcano in 1886.

Energy Masses

Wherever one is able to examine the cause and effect of every type of geological phenomena there emerges a clear picture of inexhaustible reserves of energy embodied in the earth and its atmosphere. Boreholes 2 to 3 miles down can encounter pools of oil and gas which, if not carefully controlled, can explode and become a raging column of fire. Deep mines in Mysore have working faces in gold-bearing quartz veins which explode when confining rock pressures are too suddenly released. But it is in the slow moving geological events, the

Basalts poured out over Columbia, U.S.A. Samples under the microscope show the way in which the laths of felspar enclosing brown plates of angite **4a** crystallized on cooling. Gas bubbles **4b** finally become filled with pure crystals of semi-precious minerals.

so-called tectonics of the earth, that the real magnitude of energy is displayed. At the rate of inches every year whole continents the size of South America or Australia have drifted into their present positions. At similar rates masses of strata have been squeezed out of basins to form mountain chains as continuous as the Alps or the Himalayas. Of similar dimensions are the oceanic deeps created by crustal down-warps of more than 7,000 metres (23,000 feet). These are commonplace in the Pacific near the shores of island arcs such as the Aleutians or those of Japan and the Philippines. The Indian Ocean contains the Sundra Trench to the south of Java which, like others, is a down-warp on the convex oceanward side of an arched chain of islands or a mainland range. In the Atlantic is the Puerto Rico Trench and the South Sandwich Trench each of which are over 8,000 metres deep.

The water of the ocean is also a source of inexhaustible energy. Its movements are now reasonably well charted and although masked by the violence of surface currents it moves as a body a mile or so a day and exerts an influence on the stability of the crust above and below sea level.

The energy of the earth can thus be summarized as being distributed within its three major zones – the solid crust, the mantle and the core. The mantle accounts for about 70 per cent of the entire mass of our planet and as such is the main storehouse of power. Although the mantle is about 1,900 miles thick it is only in the top 60 miles that energy is released in violent quantities. It is within this zone that the volcanoes are fed from magmatic hearths. Beneath this seems to be a layer in relative repose, which has been named after its discoverer, the 'Gutenberg Layer'. Yet another nameless layer seems to occur before we get to the layer named after Golitsyn. This is the layer from which the deep-focus earthquakes originate. The properties of even deeper layers have yet to be investigated and scientists will continue to dispute each other's interpretation of how they wrap around the core. Is the core solid or is it liquid? All seem to agree that it is composed of heavy metals and especially of iron and nickel. One thing is clear: the core is a sharply defined body possessing liquid and solid properties whose spherical outline has been revealed by seismic recordings. Perhaps it is simply a coincidence that the surface of the core has an area of almost 57 million square miles which is nearly identical with the total area of all the continents. On the other hand, the theories which are now developing in the realm of plate tectonics may find that these dimensions are connected with the origin and growth of the crust and the earth's heavy core. What will be eventually discovered will reveal in a more precise form the source and enormous magnitude of the earth's internal energy.

The craters of the moon—still the most controversial structures on its surface. The deep shadowed craters are possibly volcanic; the others may have been caused by meteorites.

Chapter Four

THE RE-CREATING EARTH

A World is Born

Speculating on the origin of the earth and its place in our solar system has been one of man's oldest intellectual pre-occupations. When man first stepped onto the moon, however, many of our long-cherished theories went by the board; for one of the greatest surprises that the American astronauts brought back to us was how similar in age are the rocks of the moon to those on earth. By analyzing them we now know that they were both formed about 500 million years ago. Meteorites – fragments of planets which have exploded in outer space – also date back roughly to the same era. From these facts we can assume that our solar system as a whole was generated around the sun about 5,000 million years ago.

The earth and the moon may be the same age, but they differ in several major characteristics. The earth is much heavier than the moon owing to its core – believed by scientists to be of nickel and iron. There does not appear to be a dense, heavy lunar core although this does not necessarily mean the moon has no core at all. Future study may well prove that it has and the discrepancy in weight between the two planets is due to some other factor.

Apart from the weight difference there are other features which are equally perplexing. If you look at a full moon you will notice that the shapes on its surface are strongly reminiscent of those which characterize the surface of the earth. The moon has mountains and flat open spaces. Similar open spaces on earth are taken up by oceans – yet on the moon there is no water. Why then are the mountains of the moon as vivid as those on earth. How were they carved out? Our mountains have been sculpted mainly by wind and rain, river-flow and ice-flow. Yet none of these exist on the moon. What kind of lunar erosion then could duplicate the landscape of the earth? It will take more than the United States' Apollo missions to find out.

We have known for a very long time that the surface of the moon is pitted with numerous craters which were thought to be the result of volcanic eruptions. But the astronauts have yet to find an active lunar volcano even though there is evidence to suggest that lavas have been poured out from time to time on the surface of the moon. These eruptions, however, would not explain the vast number of craters visible through radio telescopes.

We have craters on earth similar to those on the moon such as the meteorite crater of Cañon Diablo, Arizona. This crater was formed when a large meteorite struck the Arizona desert some 40,000 years ago. It buried itself 1,400 feet into the ground and gouged out a crater about a mile across and 600 feet deep. In every respect it resembles many of the craters on the moon.

For the moment then we must assume that the multitude of craters on the moon were produced by meteorites bombarding its surface from outer space. Yet this would not explain why rift valleys and canyons on the moon are identical to those on earth when there is no sign of water or erosion! So far the moon has kept its secrets.

The origin of the earth is especially shrouded in mystery. It seems reasonable to assume that our planet began as particles of mineral matter condensing in space to form a spherical mass. The mass would be a coagulation of gaseous and solid matter giving off immense heat. As time went by, spinning in space, it began to cool. The process of cooling may have started first at the surface and so formed a hard crust under which the particles began to further condense to yield eventually a core of nickel and iron. On the other hand, these particles could well have separated out and the heavier ones (composed of nickel and iron) condensed to form the core and then further particles condensed layer by layer around its surface to form the earth's interior. Finally, the crust was formed. But in its formation, by some process or another, water was created and thus

The most expensive rock in the world! A sample of moonrock brought back by the Apollo 12 astronauts—a dense lunar basalt containing a sample of the magnetic field of the moon.

Meteorites **(see inset right)** have bombarded the earth from outer space and created structures like the Sunset Crater of Arizona, U.S.A.

enveloped the crust with an atmosphere containing moisture. This watery veil was to become the most significant feature of our planet. Once the crust of the earth had been formed, the great hollows were filled with water as rain descended from the atmosphere on to the cooling earth. These hollows became the world's oceans. The evidence at the moment is rather diffuse but it appears that throughout

Astronauts of Apollo 16 exploring the dust and rocks of the moon's surface in search of water.

geological time the greater part of the earth was always covered by water, just as it is today.

As far as we know no other planet shares the earth's atmosphere. No other planet, therefore, has the form of life which exists on earth. How this life originated we have discussed in Chapter 1, but without water life would never have begun let alone have been sustained. Hence the minute investigation of the lunar rocks to see if the moon once had water in any form. Certainly there is no water in the moon's atmosphere and with equal certainty there is no water on its surface. But the astronauts sank boreholes hopefully looking for traces of water underneath its surface. Whether they will eventually discover water has yet to be seen; up to now the only traces appear to be the kind of alterations in rocks which water produces.

Wandering Continents

If you look at a map of the world and fit together the coastlines of continents adjacent to each other, you will find that they slot in remarkably well. Even the apparently ill-fitting pieces can now be explained. The gradual realization that continental coastlines are not accidental led geologists to the conclusion that these land masses must have split and drifted apart. This process we now refer to as 'Continental Drift' or more graphically 'The Wandering Continents'.

For a long time we have realized that the coast of South America dovetailed into the coast of Africa. In addition to the geographical fit the rocks were also found to be identical both in character and age. For instance, the rocks running towards the coast of Ghana and Nigeria match up both in age and character with those across the Atlantic of the east coast of Brazil. In Ghana the oldest rocks approaching the coast are 200,000 million years old, while in Nigeria the oldest rocks go back only 600 million years. Therefore a line divides these two groups of rocks. More significantly it was found that the jig-saw pieces fitted much more snugly if the outlines of the continental shelf on either side of the Atlantic were brought together. This evidence seems to prove conclusively that once upon a time the earth was composed of two huge landmasses which broke up and drifted apart.

Perhaps the most striking evidence of continental drift is provided by the deposits of coal which were formed in the geological period, Permo-Carboniferous, 230–300 million years ago. Within this time-span huge forests deposited vast beds of humus. From these layers seams of coal were eventually formed by becoming compressed beneath successive deposits of sands and muds. An examination of the fossil plants found in the coals of Europe has shown them to be identical with those found in the Carboniferous coalfields of North America and Central Russia. This points strongly to the existence of a forest which stretched from the Atlantic to what is now central Russia. Since the trees and plants which produced these layers of coal could only have survived in equatorial regions we can assume that these coal-forming jungles were connected by land bridges sited along the equator – 1,000 miles or more to the north of its present position! If this were so then either the equator or the land masses have changed position over the last 200 million years.

Some scientists have contested the theory of continental drift because it was difficult to visualize large continental masses like North America and South America drifting westwards away from Europe and Africa. It was also difficult to picture the substance on which these continental masses could literally float apart. The answer lies in the ocean floor. Recent submarine exploration has revealed that the ocean floor is as vividly contoured into mountains and valleys like the surfaces of the continents. But of more striking character has been the discovery of long mountain ranges which run down the central areas of the oceans. These have been called the 'Oceanic Ridges'. They are peculiar in that they are formed almost entirely of igneous rock – rock formed from molten lava. These long sinuous ridges have been developed by extrusions of lava from the earth's interior and they lie almost equidistant from the neighbouring continental shelves and are now known to be the line of split along which the oceans have been expanding. From this data geologists have concluded that the earth's crust consists roughly of *six* major masses of rock which are steadily being forced apart by this wedge of molten rock along the ocean floor. These masses of rocks have earned the name of 'plates' and the movement which they indicate is referred to as 'tectonics'. So we now regard the earth's surface as being composed of six tectonic plates on which rest the continental masses rising above sea level.

Expanding Sea Floor

Surprisingly it is not the crust as a whole which is expanding to make way for the position which the continents now occupy – but the floor of the ocean. This so-called sea-floor spreading along the oceanic ridges explains many features which have puzzled scientists for many years; for example, why earthquakes are particularly prevalent in the central areas of the oceans. It explains why islands of volcanic material suddenly appear in the Pacific: because lava is being squirted out from the earth's interior along one of these rifts in the ocean floor. And it explains why the area around Greenland and Iceland is eruption prone: because there is a split which takes place between the plates governing the movement of North America and the movement of Europe.

The scientific work into plate tectonics has reached a sophisticated stage where it is now possible to calculate how long it has taken for the Americas to move westwards, for India to leave Africa and attach itself to Asia, and for Australia to move eastwards into its present position. Calculations have shown that within the Pacific Ocean, which is about 10,000 miles wide, continents like Australia could have drifted away from Africa during the past 100 million

62

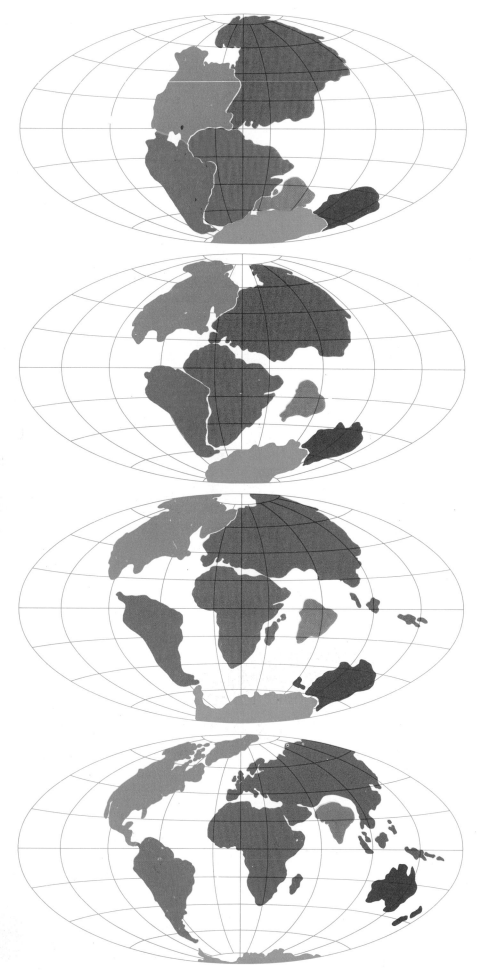

(top right) A perfect specimen of a petrified plant embedded in mud during the Carboniferous Period.

(bottom right) The bark of a tree fossilised over 250 million years ago.

Pangea: the vast land mass as it was over 200 million years ago.

About 135 million years ago the rifts of the ocean floors had forced continental masses into these relative positions.

About 65 million years ago the continental masses had drifted to these positions.

About 40 million years ago land masses like India and Australia were close to their present-day locations.

years. The movement of India into its present position was more rapid than this, but generally speaking it would appear that this great oceanic spread in the Atlantic and Pacific took place during the past 250 million years. While this may seem a long time it is comparatively short in terms of the birth and formation of the land-masses and oceans of the world. The world or crusted globe has been in existence, more or less as it is today for the past 4,000 million years.

A Radioactive Clock

None of the above conclusions could have been reached without scientists having the means to determine precisely when rocks were formed. We can now calculate when a rock was formed from a molten state with considerable accuracy even though this may go back thousands of millions of years. The technique involves the use of what might be described as a radioactive clock. The clock works on the principle that when a radioactive substance is formed it immediately begins to give out energy and form itself into new substances called isotopes. This so-called process of decay becomes a measure of time. For example, if a mineral contains uranium, the moment it is consolidated from the molten condition it starts to decay to form lead. Now this lead is not like ordinary lead; it is said to be an isotope of lead, formed by radioactivity. By calculating the quantity of radioactive lead to uranium in the rock we are able to tell how much time has elapsed to form the isotope. Similarly, if a mineral contains an isotope of potassium, it immediately begins to decay to

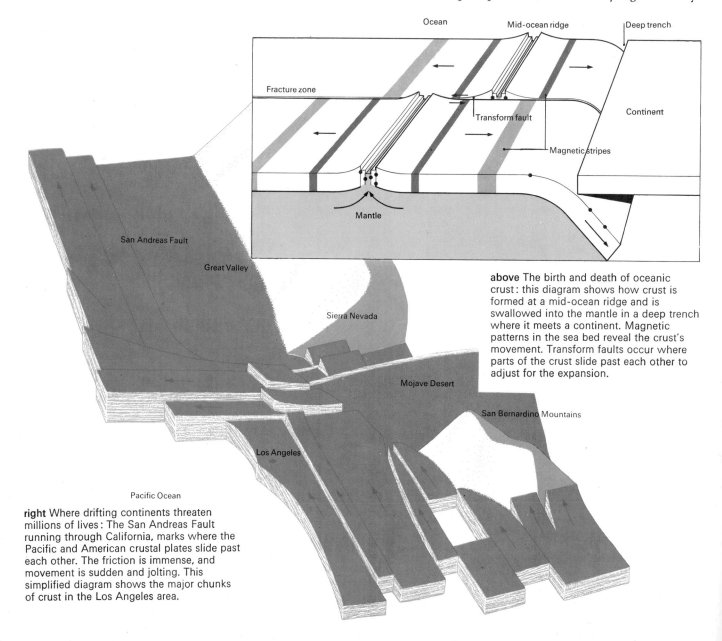

above The birth and death of oceanic crust: this diagram shows how crust is formed at a mid-ocean ridge and is swallowed into the mantle in a deep trench where it meets a continent. Magnetic patterns in the sea bed reveal the crust's movement. Transform faults occur where parts of the crust slide past each other to adjust for the expansion.

right Where drifting continents threaten millions of lives: The San Andreas Fault running through California, marks where the Pacific and American crustal plates slide past each other. The friction is immense, and movement is sudden and jolting. This simplified diagram shows the major chunks of crust in the Los Angeles area.

form calcium and the gas called argon. The argon is retained in the mineral and so by calculating the volume of argon and the remaining amount of undecayed potassium we have a measure of time.

Thus by sinking boreholes into the ocean floor, the rock samples tell us at what period of time it was formed. From the surveys already carried out we have learnt that the whole of the Atlantic and the greater part of the Pacific did not exist over 250 million years ago. This area has developed as the result of the break-up of a vast mass of rock which geologists refer to as Pangea. It was the Pangea which broke up into the six major plates and these drifted apart to give the shapes and sizes of the continents we know today. Over a period of 250 million years it would appear that the six continents have been drifting apart at an average rate of one centimetre a year. The most rapid of these movements so far recorded is along the east Pacific Rise and the slowest seems to be on either side of the Mid-Atlantic Ridge. The Atlantic is by no means as violent a submarine area as the Pacific and this is borne out by the fact that earthquakes and vol-canoes abound in and around that vast ocean. It is only towards the northern and southern ends of the Atlantic that similar volcanic activity is rife – witness the volcanic eruptions in Antarctica, those around Iceland and Greenland and towards the North Pole.

Readings have shown that it took about 200 million years to separate South America from Africa. By contrast, however, the separation of Australia from Antarctica was probably achieved in about 40 million years.

Obviously these tectonic plates cannot move without bumping into one another and at the points of collision the leading edge of one dives down under the trailing edge of another. At these points we find that the floor of the ocean has been depressed to form submarine trenches sometimes many miles deep (see opposite). In addition to the sucking down of the edges of these plates, there is a re-forming process which is necessary for the mechanism of ocean spreading to persist. As the forward edge of the plate dives down into the depths of the ocean it encounters tremendous heat and it is thus converted into molten rock again. This molten rock serves to replenish the magma of the earth's interior. Magma is the material which is exuded through the cracks in the Mid-Oceanic areas, that is, along the oceanic ridges, and it clearly must be replenished or the earth would shrink. For our planet to retain its original volume this regenerating process must be the mechanism by which the continents are being continually forced apart.

The Ocean Floor

The examination of the earth's oceanic floors has also revealed some other remarkable features. The most important of these is the altera-tion which has taken place from time to time in the magnetic field of the earth. We are accustomed to thinking of a magnetic field as always being orientated towards the north pole. But it is more com-

plicated than that. The magnetic field on either side of these oceanic ridges suddenly reverses itself so that the north pole in one strip becomes the south pole in another. This reversal of magnetism has provided the means of plotting the structures which lie on either side of the ridges and it is a remarkable fact that the pattern on both sides is identical.

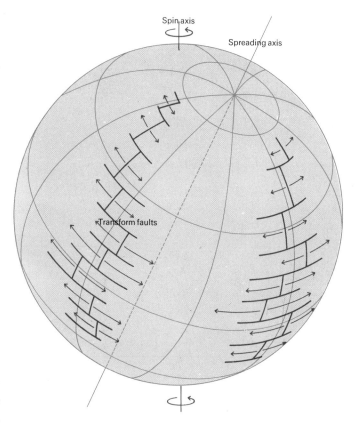

When rigid plates move in relation to each other across the surface of a sphere such as our planet, their direction and rate of movement follow simple mathematical rules. There are on the surface of the sphere two 'poles of spreading', and an 'axis of spreading' passes through them. These are quite separate from the axis of rotation and its poles – which we call the North Pole and the South Pole. If imaginary lines of latitude and longitude are drawn in relation to the poles of spreading, plates always move so that the transform faults run along the lines of latitude. Movement is slowest near to either pole and most rapid far away from them – that is, at the 'equator of spreading'.

In the light of this new knowledge it would be naïve for us to regard the earth as a straightforward linear magnet. The reversal of the magnetic fields calls for a more erudite explanation. The most plausible is that the earth behaves like a dynamo which suddenly reverses itself and so changes the magnetic field. Surveys have shown that the magnetic properties of sea-floor rocks alternate in their direction and these periods of reversal prevail for set lengths of time.

By geological standards these reversals have been relatively rapid. Detailed records of rocks whose ages range back to 76 million years, have revealed over 171 reversals of the earth's magnetic field. The earth, as it is at the moment, seems to have been in the normal condition of magnetism for the past 700,000 years. This is somewhat exceptional and we are due for a reversal although we cannot predict when. The enormous strains and stresses involved in a reversal of the earth's magnetic field creates increased earthquake and volcanic activity.

In essence, the picture presented by plate tectonics rests upon two assumptions. The first is that new crust is always being formed at the axis of the Mid-Oceanic Ridges forcing the plates to move sideways. The second is that the frontal edges of the plates dive under the trailing edges of the forerunning plates and descend and melt into the crust of the earth. In this way the underlying mantle, which is potentially molten rock, is constantly being replenished. The concept of plate tectonics depends upon this continuing formation of new crust so that the earth remains virtually constant in volume. We can only prove that the underlying mantle is being re-charged with melted down rock by deep-sea drilling. America and Russia, with other interested countries, are already considering the setting up of an international organization for this purpose. To drill through the crust on the ocean floor is going to be as exciting as man's first landing on the moon. It will be as dramatic as the tales of Jules Verne since we know not what to expect once we penetrate the solid floor of the ocean to depths of up to ten miles. Such depths are well within the reach of floating oil rigs such as those prospecting for natural gas in the North Sea. The results of this exploration will not be entirely academic: they will enable man for the first time to tap the enormous sources of thermal energy which exist beneath the solid crust.

The Oceans of the World

To understand how our earth came to be shaped as it is today we must appreciate the significance of the oceans and the events which have taken place on the ocean floor. One of the most remarkable features of the oceans has been the constant volume of water which has occupied the surface of the earth throughout geological time. The level of the sea has varied from time to time; during the Ice Ages, for instance, great volumes of water have been locked up into masses of ice in the Polar regions. These Polar regions were not always situated in their present position, and this withdrawal of water from the

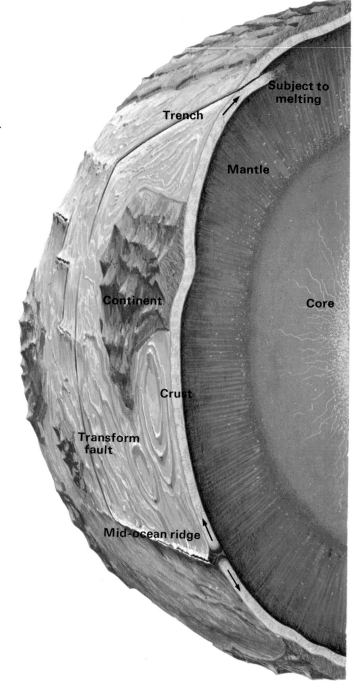

A cross-section of the Earth's Crust showing the mechanism of ocean floor expansion created by submarine eruptions forcing the plate-like crust apart and the cracking of it with transform faults.

oceans into ice has frequently led to geological events of great interest. For example, during the recent Great Ice Ages the continents were stripped of water and the continental shelves of Europe and North America were laid bare.

At no time has the sea completely covered the land although it has advanced over coastal plains at different times. There must be some mechanism that maintains a kind of equilibrium, otherwise water expelled from the earth's interior by volcanic eruptions would have swelled the size of the sea; but this has not occurred. It could also be expected that water would evaporate into the atmosphere and so reduce the amount of water on land, but this also has not occurred. By a process of continuous replenishment the atmosphere releases water as rain and appears to have maintained a balance of dry land and water throughout the life history of the earth. As a result, practically every event which has led to the formation of the surface architecture of the continents has been controlled by this replenishing process. Rains, rivers, ice and the tidal action of the seas have been the most powerful of the natural agents of erosion as well as the agents of construction of new lands out of old – by the formation of sediments.

The first important factor which emerged from ocean floor exploration was that the oceans are quite different from the continents. The mountains buried beneath the seas are nothing like the Alps, the Rockies or the Himalayas, which are built up out of folded, sedimentary rocks. Beneath the oceans there are world-encircling mountain ranges which lie mid-way between the continental masses built up from lavas like basalts. These have been forced outwards into the centre part of these ridges from the earth's interior. In places, however, these undersea mountains are covered by sediments which have been washed into the oceans by rivers from the continents.

One interesting feature of the oceanic ridges is that they have along their axis great crack-like valleys from the floors of which lava is being constantly ejected. In this way these valleys act as a wedge forcing the rocks on either side to move continuously apart. This is the mechanism of sea-floor spreading. As the rocks on either side of the ridge move apart, new rock solidifies in the crack and this in turn is broken by further extrusions of lava emerging from great depths beneath. The rate at which these cracks open up determines the speed at which the plates move sideways. The fastest rate is the spreading which takes place from the east Pacific Rise and the slowest are those which expand the Mid-Atlantic Ridge. The rate of production of new crustal rock in the central valleys of these ridges shows the speed of movement – as much as 16 centimetres (6 inches) per year. Such movements, geologically speaking, are very rapid and it gives some idea of the time which has elapsed during which some of these oceans have been formed. For example, the Pacific Ocean, which is about 10,000 miles wide, could have been produced within 100 million years.

Molten rock from the mantle forces its way into the cracks of the ocean floor and pushes the continental plates sideways.

The spreading of the ocean floor is facilitated by the formation of faults along which fractured plates will slide. These are called 'transform faults' and they present the most positive evidence for the directions which the continents are following. The movements of, say, the Americas are strongly influenced by the development of these transform faults. Areas along these transform faults are earthquake-prone; the best example is the San Andreas Fault which has wrought violence on California over a long period of time and will continue to do so.

In the same way as rocks move sideways on the ocean floor, so faults on the continental surfaces achieve similar results. If you look at the coastline of Scotland you will observe the great indentation along the east coast marking the Firth; running southwestwards is the Caledonian Canal. This canal is situated along a fault known as the Great Glen Fault; according to the rocks on either side a movement of 50 miles took place over 300 or more million years ago. In other words, the rocks have been moved southwestwards along this transform fault by sideways movements very similar to those which have broken up the submarine ridges.

The Birth of Mountains

Having accepted the concept that the continents are huge plates floating on the mantle, we must turn our attention to the mechanism which has operated within these continental masses. Explanation must be found for the existence of huge mountain chains. An explanation must also be found for the great thicknesses of sediment which have been developed to form these chains. Reasons must be found for the way in which the rocks are broken by faults and displaced either vertically or horizontally by earth movements. We must also try to explain why volcanoes erupt in certain places and not in others. All these events are clearly produced by the generation of forces underneath the continents expressing themselves upwards.

The surface of the continents is dominated by the presence of rocks

The San Andreas Fault viewed from the air over California (left).

The internal forces which have thrust huge wedges of crumpled strata in the erection of the Himalayas.

formed from sediments which were laid down in the sea. From this simple fact we must assume that they were accumulated in conjunction with ocean floor spreading. Since the Pacific and Atlantic Oceans were formed 250 million years ago the sedimentary rocks which are of greater age must have been developed on the continental shelves of pre-existing plates.

This seems to have been the mechanism by which the surface of the earth has been built up throughout geological time. The continental shelves of the land masses of any geological period must have received great volumes of sediment eroded from the exposed surfaces. These were deposited in the sea and incorporated within themselves the life of that period which has since become fossilized. By the study of these fossiliferous deposits we can now date the build-up of the rocks about us. This is a complicated pattern since it has been affected by the repeated collision of one plate after another buckling up the horizontal rocks of the continental shelf. The result was upfolds and downfolds – what geologists call 'anticlines' and 'synclines'. In many places these continental shelf areas subsided as much as four to five miles and the sediments of a particular geological period accumulated on this subsiding surface. These subsiding zones of the sea are called 'geosynclines'. When they are compressed by the sideways movement of a plate processing itself across the ocean floor, they become buckled and folded; and what was a hollow of immense depth filled with sediments, now becomes a mountain chain forced up concertina-fashion above sea level. In this way the Alps and the Himalayas were born.

Within the core of these uprising masses the molten material of

The mechanism of ocean floor expansion created by submarine eruptions of pillow lavas forcing the plate-like crust apart and cracking it with transform faults.

As the continents are forced aside by the movement of the crust they form geosynclines which sag to depths of up to 20 to 30,000 feet. The sagging effect causes the sides of the geosynclines to crack and along these lines volcanoes erupt. The lava mixes with the sediments on the flanks of the volcanoes and is washed into the sea by the uprising land masses.

the earth's interior rose also and cooled to form the rocks we know as granite. So wherever erosion has removed a mountain and revealed its heart we almost invariably find a core of granite. By plotting these granite cores we can trace the axis of mountain chains which have long since been erased by erosion. The granites of Cornwall, for example, date back some 200 million years and match up with those in Ireland and on the continent as far as the Hartz mountains in Germany.

In the distant geological past geosynclines have frequently led to the accumulation of sediments containing great mineral wealth. Over 600 million years ago the southern flanks of ancient Africa began to subside. Into this geosynclinal hollow the sea poured vast quantities of sediment eroded from the crystalline rock of the interior. Among these rocks were veins of gold. As gold resists erosion the particles were swept by tidal currents onto the foreshore of the ancient sea and there they concentrated themselves in the shingle and sand of the shorelines. Further subsidence caused them to be covered by muds and so the process of subsidence protected them for mankind to recover in our own times. This, broadly speaking, was how the two-mile deep basin of the Rand was formed.

Such geosynclinal depressions have taken place many, many times in the past and most of the mountain areas like the Urals, the Rocky Mountains and the Andes have been formed by the compression of rocks originally laid down as flood-like sediments in areas invaded by the spreading oceans. On the flanks of these subsiding areas the rocks were naturally in a state of tension; and it is along these tensional margins that volcanoes erupted and produced lava. Along the coastline of Pembrokeshire in Wales the product of such volcanic eruptions is clearly visible on the margin of the geosyncline. To the west of Fishguard Harbour there are over 2,000 feet of lavas which were poured out over the sea floor. Three miles or so inland there was a volcano erupting which created cones 4,000–5,000 feet above sea level. That happened over 500 million years ago, but the same thing occurs today in areas like Iceland which is situated in a marginal zone on the fringe of sedimentary basins possessing the oils and gas of the North Sea.

Rivers as Sculptors

A clear understanding of the simplest of natural processes can often reveal the drama of events which have taken place in the past. The history of any river will show that its waters have eroded its channel against an uplifting mass of rock. If this continues for a sufficiently long time the river cuts a gorge and if it flows through a desert area it erodes a canyon. Even a superficial glance at the awe-inspiring flanks of the Grand Canyon of Arizona shows that the river has cut its way through stratified pages of geological history to a depth of nearly a mile, covering a period of 300 million years. At the bottom of the canyon there are hard crystalline rocks which were formed when that part of Arizona was a plate. On this plate these sediments have been accumulated by the process of subsidence. Then the process was reversed and these rocks have been forced upwards with the Colorado River contesting this rebellion of the earth's interior.

By a similar process other great canyons were formed, like the one which descends from the Himalayas between Mount Everest and

Kangchenjunga. This enormous gorge cut by the River Arun could not have been achieved unless the whole of the Himalayan area had risen something like 6,000 feet. It is this kind of elevating process which has given that vast chain of mountains their gigantic stature.

Canyons are not formed in arid regions only: the basaltic plateau of Columbia in North America was carved by glacial waters. As the ice cap from the north spread southwards into the valley through which the Columbia River had been flowing westwards to Vancouver, a huge lake developed and its waters overflowed bearing with them sand and mud from the glaciers. This water ate its way rapidly through the basalts and dissected them into deep canyons over 1,000 feet deep. At one place, appropriately called Dry Falls, a waterfall was created – over 400 feet high and three miles wide. When it was in full spate it must have thundered with the roar of a hundred Niagaras combined. But now no water falls over its brink. It became Dry Falls when the ice front receded and the Columbia River reverted to its old course and drained the lake. Franklin D. Roosevelt was made aware of this piece of geological history and from it emerged his imaginative development of the Coulee Dam. He ordered a dam to take the place of the ice front and so again created the lake whose overflow water is now used to irrigate the waste lands of the Grand Coulee and to create vast quantities of hydro-electricity. This power and water have resulted in the production of millions of acres of exceedingly fertile land from the barren Coulee area – an interesting example of how man can copy nature with highly satisfactory results.

Future of the Continents

Volcanic eruptions, earthquakes and the process of ocean-floor spreading never cease. The crust is as restless now as it has always been. Dr G. S. Pawley and Dr N. Abrhamsen have recently shown that even the pyramids of Giza have probably moved. They argue that the pyramid builders intended that these tombs should have their walls running north-south to east-west. Assuming this to be true they point out that the alignment of the pyramids is now four minutes off arc to the west of north. Since there are no cracks which would indicate that earthquake faults had moved them, the only explanation seems to be that Africa itself has moved in the last 4,500 years.

The change in the alignment of the pyramids bears out the fact that earth masses are constantly on the move, and this discovery is a neat expression of the rate at which rotations of this kind are taking place. It also adds further evidence to support the view that millions of years hence the Red Sea will be cut off by a plate which will break away from the east coast of Africa and drift northwards as the continent continues to rotate westwards. This series of projected movements has been carefully evaluated and it will leave the world in the condition shown on page 74. Geologists now believe that

The Grand Coulee Dam – a concrete wall containing hydro-electric turbines erected along the line which was dammed by ice from which the flood waves of the Ice Age scoured valleys and canyons throughout Columbia.
The pyramids of Giza. Movements of the earth's crust are suggested by their realignment.

The Grand Canyon eroded by the Colorado River. The land rose over a mile relative to its original sea level (left).

within the next 50 million years the Atlantic and particularly the South Atlantic Ocean will grow at the expense of the Pacific. The Indian Ocean will also increase at the expense of the Pacific as a result of the movement of Australasia northwards. This will further intensify the instability of Japan and the forelands of Eurasia which lie in the path of the Australian plate. As the North Atlantic Ocean spreads wider, Europe will be forced eastwards and along the west coast of North America slivers of the continental mass will drift north-westwards. It is estimated in about 10 million years Los Angeles will be abreast of San Francisco as one such fragment slides along the San Andreas Fault. Continuing to move it is further estimated that Los Angeles will then begin to disappear like Atlantis by being sucked down into the Aleutian Trench. But do not worry if you are planning a holiday there in the near future: this will take at least another 60 million years to come about.

Antarctica would appear to be immune from such movement during this time and much of central Russia will also remain some-what static. So in the shaping of the world of the future it seems that Nature will not depart from its well-established mechanism: main-taining a crust dominated by ocean floors. The earth will continue to spin more or less on its present axis and allow the continental plates to rise and fall to make new lands out of old as they always have done.

In the swamps of the tropical areas of the earth about 100 million years ago enormous monsters roamed and lived as vegetarians. Stegosaurus, 25 feet long and armoured with back plates, lumbers in the wake of Diplodocus, the largest of the dinosaurs measuring 90 feet from head to tail.

The globe of the future. With the continual movement of land masses the world will look something like this in 50 million years' time.

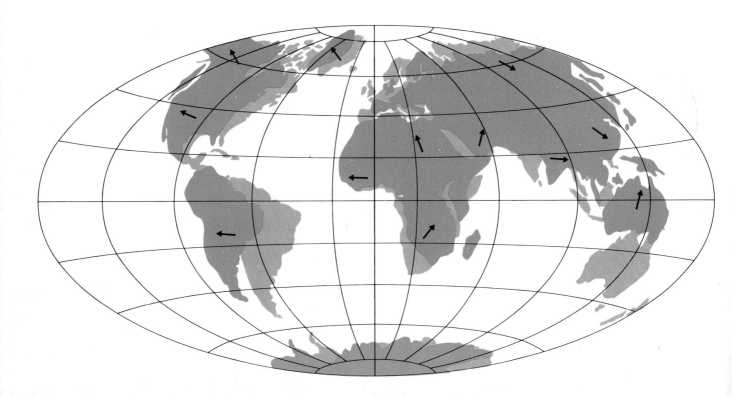

Chapter Five

FOSSILS: THE MUSEUM OF LIFE

We have only one sure way of detecting the pattern of life as it has developed on earth: the study of fossils. They alone reach back across the ages to reveal the trends which enabled Darwin and Huxley to build up their concepts of evolution. There is no fact in the great ocean of knowledge more wonderful than the speck into which Nature has packed the secret of life—a speck too small for the eye to see. The living cell is something no man yet understands but from it has originated the creatures of the earth and made them different from stone. It is equally true that man does not yet know how this earth originally developed a crust of stone, but like the living cell it must have grown from invisible atoms and molecules, and having begun it has continously been subject to processes of growth and decay not dissimilar to living matter.

Minerals may have great material value but the hunt for fossils in rocks is a pursuit which can provide a source of intellectual excitement no money can buy. This is why some of the most exciting finds have been made by so-called amateurs who love hammering at rocks in search of fossils. It is useless searching the rocks produced by the earth's interior for fossils as magma is too hot to allow organisms to exist or be preserved. So it is to those rocks formed from sediments that we turn to find fossils and in particular those which have been laid down in lakes, deltas and the open sea. Most of these sedimentary rocks contain fossils but these are not always easy to recognize; very often they are the remains of animals and plants which do not exist today.

When fossils were arranged in an orderly fashion in the middle of the last century, it was found that they grouped themselves into fairly well-defined time zones similar to the stages in the development of man's life on earth. It was also found that these different groups of fossils were embedded in the layers of sedimentary rocks like the print on the pages of a history book. We are now able to divide the sedimentary rocks into ages which have been given historical names.

Those fossils which marked the earliest period in evolution as belonging to the earliest forms of life are called Palaeozoic (Palaios: ancient). Then there were those which represented life between ancient and modern times which were called Mesozoic (Mesos: middle). Finally, the latest fossils are referred to as Cainozoic ('Cenos': 'recent'; 'zoe': 'life'). Within these three broad eras of time geologists have found that the pages of rock-history contained divisions which subdivided into well-defined stages in the evolution of life as we know it in its complicated present-day pattern. But throughout the past 600 million years or so the records of life on earth have been preserved in sediments which were largely developed on the lowlands or the continental shelves of the continents as they existed in each period of time. This incidentally is why rocks of the continental shelf now contain oil – the product of decay of these buried organisms as they became fossilized. Now that we have discovered the mechanism by which the land masses have developed over the past 250 million years, it is possible to gain a much clearer picture of the distribution of land and sea throughout the previous 400 million years when ancient or Palaeozoic forms of life were emerging from one ancestral stock or another.

The Palaeozoic Period

Schematically each period of geological time has been accorded a name which has been based either on the place the fossils of that age were first found or on some outstanding characteristic of the deposits in which they occur. Thus, the oldest age group belongs to the Cambrian Period because fossils and rocks containing the earliest fossil remains were found in that area of Wales which was occupied by ancient Britons called Cambraes. The next period was named Ordovician after the tribe of Ordovices; and this was followed by the Silurian Period after the Silures. The next period was classified as either being Devonian or Old Red Sandstone. This dual terminology referred to the marine deposits first identified in Devon at a time when sandstones and marls were being accumulated on land under desert conditions. Finally, the Palaeozoic Era ended with the development of deposits which were heavily composed of carbon compounds and for this reason it was called the Carboniferous Period.

The Mesozoic Era

These generic titles, first named by British geologists, were so clearly defined that it became possible to correlate deposits under such historical headings all over the world. The same has been true of the names used for the periods of time which make up the Mesozoic Era but instead of being exclusively derived from the British Isles they cover areas of Europe and elsewhere.

The earliest period of the Mesozoic Era is called the Permian, named after the ancient kingdom of Permia situated east of the Volga. Next comes the Triassic Period, a name derived from a three-fold division of rocks in Germany. The Jura Mountains give their name to the succeeding Jurassic Period but this is followed by the Cretaceous Period named because it was the great chalk (creta) forming episode at the close of the Mesozoic Era.

In classifying the Cainozoic Era a somewhat different process was used. Names were coined to indicate whether the rock formations contained more or less numbers of fossilized modern shells. These are given in the geological time scale presented in semi-diagrammatic form. This pictorial representation of geological time and the evolution of life indicates that thicknesses of sediments were accumulated during each period and from this it will be seen that the crust must have sagged frequently to depths 8 miles or so in places to allow these sediments to accumulate. Out of these columns of strata great mountain chains were erected only to be eroded to lowlands. But sufficient stratigraphical evidence has survived to enable geologists to locate their former existence.

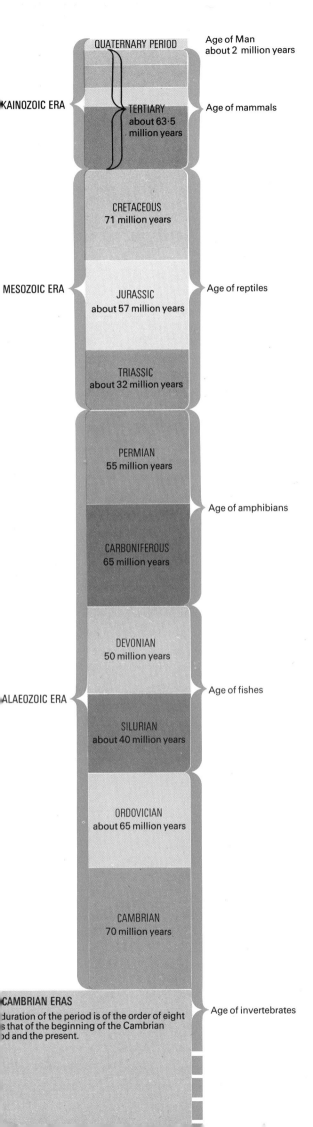

QUATERNARY PERIOD

Age of Man
about 2 million years

KAINOZOIC ERA

TERTIARY
about 63·5
million years

Age of mammals

CRETACEOUS
71 million years

MESOZOIC ERA

JURASSIC
about 57 million years

Age of reptiles

TRIASSIC
about 32 million years

PERMIAN
55 million years

Age of amphibians

CARBONIFEROUS
65 million years

DEVONIAN
50 million years

Age of fishes

PALAEOZOIC ERA

SILURIAN
about 40 million years

ORDOVICIAN
about 65 million years

CAMBRIAN
70 million years

Age of invertebrates

CAMBRIAN ERAS

duration of the period is of the order of eight
s that of the beginning of the Cambrian
d and the present.

This diagram shows the geological periods
and the evolution of life on our planet from
Pre-Cambrian times to our own era.

The Origins of Life

At the beginning, organisms were presumably simple and robust and did not seem to need a skeleton or a shell to aid them in locomotion or defence against other forms of marine life. This we know from the absence of fossilized remains in sedimentary rocks dating back beyond 800 million years. Lately, however, it has been found that these very ancient sediments preserved the remains of plankton which teemed in primaeval seas as prolifically as they do today. We have still a great deal to learn about them but it is remarkable how their delicate form has survived the years.

The first real glimpse we have of the earliest types of life is a somewhat surprising one. Instead of finding the fossilized remains of rather simple organisms, they turn out to be sophisticated in both anatomy and behaviour. As mentioned earlier the rocks containing these various types of fossilized shell fish are placed in the Cambrian Period. Quite suddenly almost every type of organism, with the exception of vertebrates and true plants, made their appearance. There were algae which developed great thicknesses of limestones; there were molluscs, jellyfish, sponges, starfish and brachiopodia. In addition to these there were remarkable organisms called trilobites and graptolites – neither of which have we any knowledge now.

As the name indicates trilobites were three lobed, crab-like organisms having a head shield, a middle lobe covered and protected by a series of articulated ribs and a tailpiece. These three-piece protective shells developed extraordinary shapes all designed for defence and locomotion at sea. They had eye pieces which resembled those of insects although how these operated under water is something of a mystery. Crawling along the shore like crabs they were capable of a kind of protracted strip tease: they could shed their shells and move away, presumably to rest or reproduce and then grow new shells as they adapted themselves to various conditions along shorelines. There is no evidence that they were capable of swimming or that they could live in fresh water so that where the trilobite fossils are found we can safely say that here was a shoreline at least 400 million years ago. As a result we can trace these Cambrian, Ordovician and Silurian shorelines all over the world and so obtain a picture of the distribution of the continental shelves prior to the break up of Pangea – the huge land mass that existed before the Atlantic and Pacific oceans.

These insect-like trilobites were exceedingly sophisticated and sometimes they could have a body length of over 12 inches. From beneath their head shield, feelers or antennae enabled the trilobite to sense changes and trap its food. How efficient were the compound eyes we cannot say but from their shape and size it is possible to gain some idea of whether they lived in clear water or the musky depths of the ocean. Behind the head shield the body was covered by segmented flexible ribs of chitin to protect its double-branched limbs. Each limb was jointed like an insect's leg, obviously to enable it to walk rapidly over soft sand and mud. Attached to each leg there was

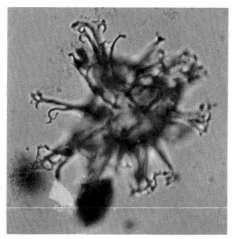

A fossil plankton still preserved in its original cover of chitin.

Perfect Trilobite (calymene) fossils about life size, found in sediments over 400 million years old. Note the prominent eyes in the head shield.

Graptolites. Fossils about life-size found in Palaeozoic rocks. These tiny marine organisms, now extinct, floated like seaweed on the water surface.

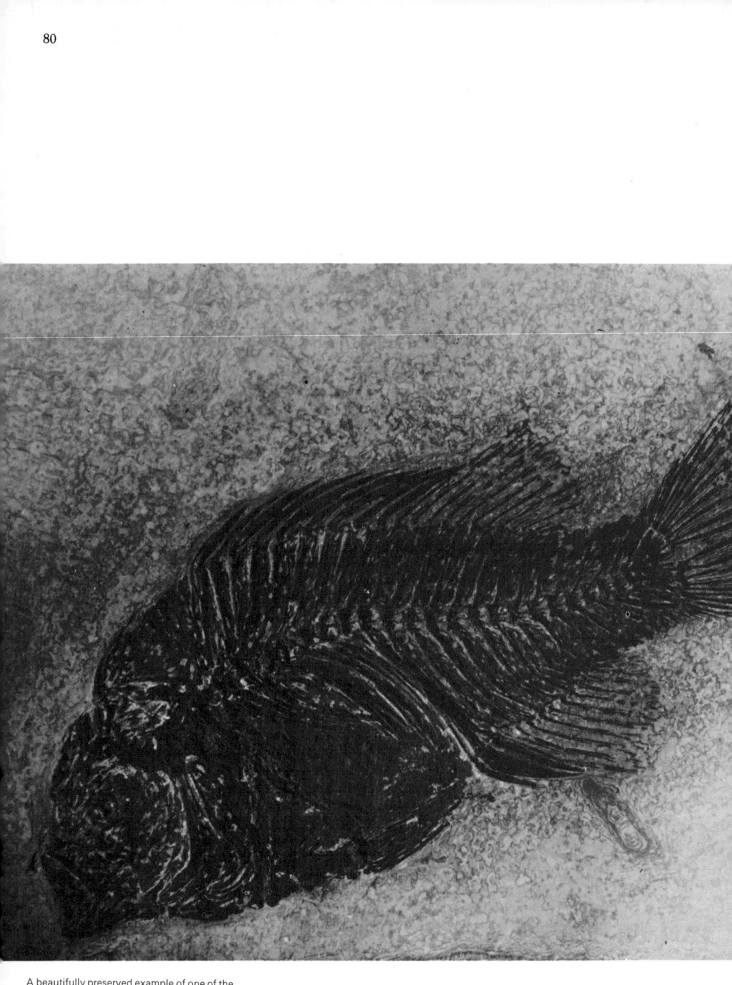

A beautifully preserved example of one of the
earliest remains of the fossil fishes.

also a feather-like appendage which was an aid to locomotion. The third lobe, or tail piece, took on a variety of shapes as is the case of most aquatic animals depending upon whether it was to be used principally as a paddle or for defence. Fossil trilobites have been found which show that they could curl up and the tail piece then helped to protect the head from attack.

The other remarkable organisms which we find in these ancient or Palaeozoic rocks are the graptolites. There is nothing quite like them on earth today. They were tiny colonial marine organisms which floated like seaweed over vast areas of those ancient seas. As fossils they resemble graphitic markings in a rock, hence their name, graptolites – derived from the Greek word 'graptos' meaning 'written' and 'lithos' 'a stone'. They are somewhat difficult fossils to spot in a rock but on detailed examination we find that they consist of tiny polyps strung along a central spine or canal extending from a central conical structure called a 'sicula'. The function of the sicula was certainly a central point of growth or reproduction of these colonies but lately it has become evident that the sicula played an important part in the locomotion of the unit and enabled it to rise or fall in the sea or move horizontally in any direction.

The earliest graptolites were like webs but these gave way to forms with four so-called stipes (Tetragraptids), then two stipes (Didymograptus) and finally a single one (Monograptus). They died out towards the end of the Silurian Period and they have never been seen since. However, during their life-time of about 200 million years, they spanned the shorelines of the earth.

Graptolite fossils were first discovered in Britain but deposits of the same age have since been found which indicate that the sea floor of that time extended throughout Europe and Scandinavia. Seas then must have extended across Greenland and into the Arctic as far as Baffin Land and across to Alaska. Graptolites and trilobites have also been found in the Rockies, in South America, in New South Wales, Australia and in parts of Russia. In other words wherever they turn up we can pinpoint the position of the continental shelf which fringed the huge land mass 'Pangea' before it broke up into the continental masses of today.

First Fish of the Sea

Shell-fish fossils of early Palaeozoic times all tell the same story. Clearly, each group endeavoured to adjust itself to its marine environment. As the graptolites and the trilobites petered out truly fish-like creatures began to make their appearance. Some of these evolved, during the Devonian and Carboniferous periods, in vast inland lakes on Pangea while others around the coast began to diversify their shapes to combat the tides and changing temperatures of the sea.

Beginning with small fishes about 4 inches long – one of which is called Tholodus – a wide range of back-boned fishes evolved. They were like tadpoles covered with tiny scales. Lacking paired fins and

A crinord and corals which formed this limestone over 300 million years ago.

A typical coral polyp (zaphrentis) showing radial development; a reef-forming coral which lived over 300 million years ago.

having mouths with no jaws they must have taken in food by suction. Later they were joined by larger swift-swimming shark-like forms. In these seas there were also large Arachnids or 'sea-scorpions'. Guided by well-developed eyes they were able to pursue and prey upon smaller creatures. Attached to the head were four pairs of legs and the end ones being stout and flattened to form paddles. The first pair of limbs was armed with pincer-like claws.

Usually these sea scorpions were quite small but some grew to as much as 5 inches long and behaved as tyrants of the deep. Towards the end of the Devonian Period they met their match in the newly evolving vertebrates. Life in the Devonian Period truly expressed Darwin's concept of the survival of the fittest. The fishes rapidly developed jaws and sharp teeth, clad themselves with scales and streamlined their bodies for increased mobility. Another line of evolution was established by marine forms, one of which is called Dipterus, which developed the ability to breathe air. This particular form was about 12 inches long. The teeth were situated in the roof of the mouth and on the inner side of the lower jaws. A dental pattern of this kind was clearly adopted to enable Dipterus to chew soft aquatic vegetation along with any small molluscs entangled in it. In these and other aspects it resembles the modern Mudfish (Ceratodus) which lives in the rivers of Central Australia. During periods of drought the mudfish has to resort to breathing air by means of a peculiarly constructed lung. This was how its ancestor Dipterus also lived over 300 million years ago.

Curious though this may be the importance of the Dipterus lies in the clues it provides in enabling us to find the ancestors to amphibia. Two footprints have been discovered dating back to this time which proves that the first quadruped walked on land during the Devonian Period. At that time the climatic conditions in Britain and Europe were tropical as is evidenced by the enormous development of red desert deposits called the Old Red Sandstone. An appreciation of this simple fact has led to the evaluation of the rocks in the southern parts of the Irish Sea. Extending westwards from St Bride's Bay in Pembrokeshire it can be shown that a valley as deep as the Grand Canyon was excavated and filled with sediment during the Devonian and succeeding Carboniferous Period. These and later geological formations are now being tapped for natural gas and oil similar to the fields established in the North Sea.

Similar Devonian deposits containing these fossil fishes formed the oil pools of Ontario and the eastern central states of America. Deposits extended southwards through Tennessee and Alabama into the Gulf of Mexico and into the northern parts of South America. During this period the fossil fishes show that Africa had been sinking – a fact which resulted in the formation of the famous Table Mountain Sandstone. Unlike the Devonian deposits already mentioned those in South Africa indicate – from fossil evidence and glacial boulder beds – that climatic conditions were cold. This is yet another proof that the polar regions have not remained in their present position throughout geological time. In Australia the Devonian Period opened up with intense volcanic activity in the Snowy River territory. Then came a period of gentle subsidence in which over 5,000 feet of coral limestones were accumulated in the southern part of New South Wales. The period closed in the Victoria territory with the development of radiolarian cherts, mudstones and volcanic lavas and ash ejectamenta.

The Carboniferous Period

We now approach the closing stages of the Palaeozic Era: the Carboniferous Period. The name was adopted because the deposits laid down then contained virtually all the coal in Britain. The rocks are subdivided into three; the lowermost division is mainly composed of limestone and is referred to as the Carboniferous limestone which forms hill country like the Mendips, the Peak or the Pennines. Only a casual glance at exposures of limestone in these hills will indicate that they were formed at times when shell fish, sea lilies and corals abounded in the sea. Classical studies of the fossils in the Mendips and especially in the cliffs of the Clifton Gorge produced an evolutionary picture of such clarity that they provide the means of identifying the same strata everywhere in the world.

The peaceful scene of tropical seas and coral reefs in which these limestones were formed was sometimes interrupted by volcanic explosions. Typical of these eruptions was the one which burst through the limestone reefs of the area we now know as the Peak District of Derbyshire.

The development of limestone almost invariably means clear waters – the best conditions for shell fish and particularly the corals which secrete the lime. Throughout Europe, North and South America we find such Carboniferous limestones. These are usually rich in fossils, particularly of corals, sea lilies as well as bivalved shells which are sometimes large and heavily ornamented, like the Productus. Again we must remember that Pangea, the huge continental mass, still existed and although exposures of Carboniferous limestone may now be thousands of miles apart they were originally part of the continental shelf of this vast land mass. So we find that fossils around Niagara Falls are similar to those in South Wales as well as those in New South Wales, Australia, and many other parts of the world as well.

The Evolution of Coral

It was from the study of the growth of coral reefs in places like the Great Barrier Reef that Darwin first began to piece together the evidence of animal evolution. Corals respond sharply to environmental changes; so we find that throughout the Carboniferous Period there was a progressive series of adaptations which created one new species after another. Some corals grew as solitary forms; these were more wisespread than the colonial types which tended to build the limestone reefs. Typical solitary forms were the Zaphrentis and another called Dibunophyllum; both exhibited the typical characteristic of such individual species in that they needed to be anchored to a firm base by little skeletons to support their weight. As they grew they secreted layer upon layer of lime, either concentrically or radially or sometimes both. The process is rather like the development of the annular rings in a tree. This type of growth has been studied in great detail in modern corals and it has been found that these growth lines build up in daily, monthly and yearly cycles indicating the age of the coral. Applying this interesting discovery to fossils scientists have found that Devonian corals have yearly cycles of 400 days instead of 365 which suggests that the earth's rotation has been slowing down.

Colonial or reef-building corals like Lithostrotian or Halysites are typical of these Carboniferous forms and wherever they occur we can assume that the seas were warm and devoid of muddy waters. These ideal conditions did not last indefinitely; eventually the clear waters became saturated with sediments which built up sand banks and mud flats which pushed the seas back. On land the humid climate created the right environment for jungle growth which brings us into the Millstone Grit episode of the Carboniferous Period. In Europe and North America large estuarine areas of mud flats and sand banks became the dominant features of the landscape making the land surface of Pangea more congenial to life. From this time we begin to find fossil plants becoming recognizable and these marked the dawn of jungles the like of which had never before been seen on earth.

The Beginning of Coal

The Carboniferous Period came to an end with the development of several thousand feet of muds and sands with intervening beds of coal. Hence the period was named Coal Measures. Different stratigraphical names were invented for this widespread series of deposits in various parts of the world but essentially the scene was the same. By describing coals in Britain we might well be talking of coalfields of the same age throughout the northern hemisphere in Europe, Russia and North America. All of these were formed under tropical conditions – the latter day counterpart of which would be the mangrove swamps around the equator. Coal seams of this age also occur in South America, South Africa and Australia, but these were developed under sub-arctic conditions which created forests somewhat like those in present-day Siberia. Both of these geographical facts again raise doubts as to the position of the poles in one geological age or another.

One thing is certain, however: whatever the climatic conditions which created it, coal is the product of decayed vegetation regardless of its age and the conditions under which the plants disintegrated. Carboniferous coals, nevertheless, occupy a special place in this picture of deposits of solid fuels deposited in the earth's crust.

The view from Thor's Cave, Staffordshire, England, of the limestone scenery dating back to Carboniferous times.

Cross-sections of fossil icerals showing the structures which reveal the way and rate of their growth.

Another mystery is why at the end of the Millstone Grit period all the low lying lands of the earth were covered with dense foliage? Compared with our contemporary botanical scene that of the late Carboniferous times was somewhat topsy-turvy. Spore-bearing plants, which are now so insignificant, dominated the scene and formed huge stately trees, while true seed-bearing plants were unknown in Carboniferous times. In other words, flowers had not yet appeared on earth. Fortunately, spore-bearing plants such as ferns and club mosses exist today side by side with seed-bearers so the life story of each group is well known. A spore resembles a grain of pollen in size and shape and it is developed, for example, in the brown spots on the back of a fern leaf. These are the spore cases or sporangia within which the source of new plants is stored. Eventually these sporangia burst releasing the tiny spores onto the ground. There they do not give directly to a new plant but to a small disc-like body which carries sex organs on its underside. From this is released the egg cells and the sperm cells. Fertilization can only take place in water, so that spore-bearing plants can only develop in wet conditions. Seeds, on the other hand, are more self-contained and fertilize as a unit without need of water. This explains the widespread occurrence of Carboniferous coal as the product of the spore-bearing plants in the swamp lands around the fringes of Pangea.

Under these conditions in Carboniferous times huge thicknesses of rotting plants must have been formed to give each bed of coal. Assuming that this plant material degenerated first into humus and then into peat, calculations have shown that it takes roughly 10 feet of humus to yield one foot of peat. By similar calculations we find that it would take about 8 feet of peat to form one foot of an average coal bed. This presents the picture of coal-forming times. Year after year carpets of rotting vegetation covering hundreds of square miles must have been formed to yield such coal seams.

There are many coal-beds of 5 feet thickness which must have been derived from 400 feet of humus. There is nothing surprising about this when we recall that peat bogs frequently contain over 100 feet of peat built up over thousands of years in geologically recent times in temperate climates. Even so, the great thicknesses of humus involved in the formation of coal shows the stature and verdance of the majestic flowerless forests of Carboniferous times. The absence of flowering plants which are involved in the formation of newer seams of coal, such as those in India or the brown coals of Europe, makes Carboniferous coal unique in composition.

The forest humus, from which the beds of coal were derived by burial beneath succeeding layers of sediment, were accumulated in dense forests of Amazonian dimensions in the northern hemisphere. In the southern hemisphere ice-fields as large as Antarctica existed and around their tundra fringe dense forests provided the foliage for coal formation. Both forests were dominated by fern-like foliage. There were trees (Gymnosperms) which resembled conifers over 100

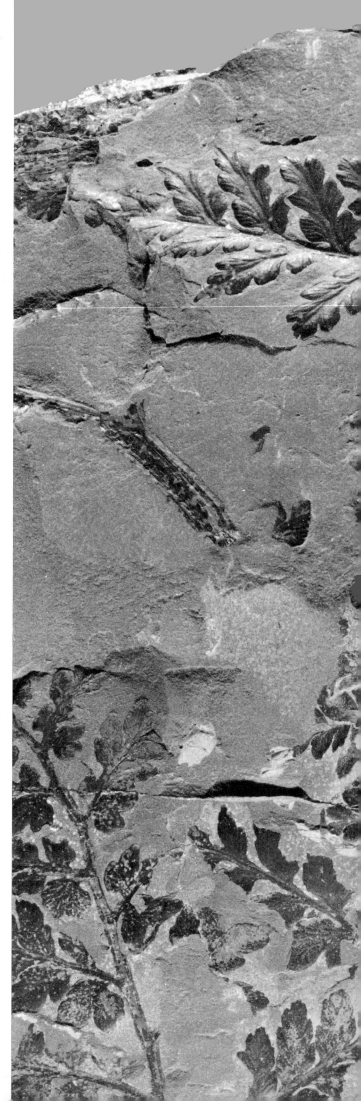

feet tall. They were like modern herbaceous plants although they grew tree-like in stature. Then there were the Lycopods. The range of such giant ferns must have been stupendous to have created so much humus. Likewise, the tiny little 'horsetail' (Equisitales) of today was a giant tree over 100 feet tall in Carboniferous times. Other plants thrived in these swamp lands. Algae must have grown abundantly in ponds; lichen occurred on the boughs of the trees and fungi grew on rotting trunks. Liverworts and mosses as well as bacteria have left their traces in the fossilized records of jungle life 300 million years ago.

At this point in time the botanical world predominated effecting major changes in biological species – changes which signalled the complexities which were to develop throughout the Mesozoic Era. The close of the Carboniferous period was the pivot about which all the characteristics of the earth, such as its crust and all living forms, began to change dramatically. We have seen how rifts opened up in Pangea to become the mid-oceanic ridges creating the Atlantic and Pacific oceans – events which must have had serious effects upon the environment.

Enter the Mammal

Certain gravel deposits which developed during the Palaeozoic Era, and especially during Cambrian times, have been found to contain grains of iron sulphide in the form of pyrite – a brassy mineral popularly called 'fools gold'. Pyrite can only be formed in the absence of oxygen so its occurrence in Palaeozoic gravels indicates that the atmosphere then was deficient in oxygen. If this was the case it has important implications – apart from the ability of animals to live without oxygen. There would, for instance, have been no blanket of ozone (a form of oxygen) in the stratosphere to shield the earth from the ultra-violet rays of the sun. At the present time there is such a zone about 10 miles up in the atmosphere which shields us from most of the sun's output of lethal ultra-violet rays. Holiday-makers know how easy it is to get sunburn. This is caused by the ultra-violet light which leaks through this blanket of ozone. (Paradoxically, although ultra-violet rays are lethal they also have the ability to promote biochemical reactions some of which are essential for living tissues to thrive, especially plant tissues.) So there is no reason why a more highly saturated atmosphere of carbon dioxide and nitrogen could have existed in Palaeozoic times.

With the incoming of plants the atmosphere could have been progressively enriched with oxygen by photosynthesis, as takes place today. Such a development might account for the sudden diversification of life on earth towards the end of Palaeozoic times. The enormous coal-forming jungles of Carboniferous times might well have been the final phase in the evolution of the earth's atmosphere and with it the proliferation of animals and plants which thrive in air and water enriched by oxygen.

A fossilized fern-like plant typical of the vegetation which fomed coal deposits over 250 million years ago.

Being mammals ourselves we are naturally interested in the origin and evolution of the vertebrates. The earliest of these remains were found in the early deposits of the Mesozoic Era. Before these were laid down the land masses were subjected to intense earth movements which resulted in a great redistribution of land and sea all over the world. For example, practically the whole of Britain and Europe was uplifted and the lands were subjected to desert erosion creating deposits which became known as the New Red Sandstone (to distinguish them from earlier desert rocks called 'Old Red Sandstone'). Eventually they were renamed Permian by Sir Roderick Murchison after he had been invited by the Czar of Russia in the early 1900s to examine the geology of eastern Russia. There Murchison found rocks similar to New Red Sandstone in the province of Perm.

The Permian period was probably one of the greatest episodes of crustal movement in the history of the earth. It was during this upheaval that mammals first appeared on earth. More positive evidence of this important step in evolution has been obtained from the succeeding Triassic and Jurassic periods. As so frequently has been the case, the earliest of these remains were discovered by amateurs interested in rocks. The first of these fossils were found in beds called Stonesfield Slate, north of Oxford, in England which were quarried for roofing purposes. A law student, W. Y. Broderip, who was interested in natural history found two small jaws and showed them to Professor Buckland of Oxford University who recognized them as being mammalian. This important discovery exploded the commonly accepted view that no mammals existed before Tertiary times, that is about 90 million years ago. Professor Buckland's conclusion received support from the great French palaeontologist Baron Cuvier who went so far as to link the bones with the primitive living mammal – the opossum. They named the fossil mammal Amphitherium which is now accepted as a mammal retaining traces of its reptillian ancestry.

The jaw, unlike that of a reptile, consisted of only one bone, and it had differentiated teeth in the form of incisors, canines and molars. Two other types of mammal rather like large rats scuttling around in the deserts were also found in the Stonesfield Slate quarries. These were called Microlestes and Phascolotherium. At an earlier period of geological time an almost complete skull of a mammal called Tritylodon was found in the Triassic rocks of South Africa. With the exception of the insectivore Amphitherium these early mammals probably resembled the pouched or marsupial mammals of today, but they may not have advanced much beyond the egg-laying or menotreme grade.

For the purposes of this book it is impossible to follow the rapid diversification of life during the Mesozoic era. Lines of development were tried by reptiles, amphibians and mammals. Some took to the air; others became vegetarian or carnivorous – each species trying desperately to achieve successful lines of evolution. Reptiles like

The flying Pterodactyl – the first attempt of animals to take to the air. An artist's impression (left) shows the vast wing span needed to keep the hefty body aloft. (below) A fossil skeleton of a Pterodactyl.

Archaeopteryx took to the air and although that particular one failed to survive it established the technique of flight which was later mastered by the incoming of birds.

At the dawn of the Mesozoic era vertebrate life was dominated by reptiles now extinct. Most of them were cold-blooded egg-layers. For example, the remains of the reptile Dimetrodon gigas in Permian deposits in Texas show it to have been like a great monitor, 9 feet long with an alligator-like head. Its backbone was extraordinary in that each vertebra had spines 3 feet long covered by a web of skin not unlike a sail. We have no idea what this vertebral structure was used for but it is believed that only the male Dimetrodons had this strange and cumbersome device.

Many of these reptiles were well adapted for life in the sea, especially the Pleisiosaurs. They had barrel-shaped bodies, a long neck and tail and limbs which were modified to act as flippers for swimming. The long neck was particularly useful for quick darting movements to snap up all kinds of fishes which they quickly macerated with their sharp teeth and digested by means of pebbles which they swallowed for use as 'stomach stones'. It was quite common for these Pleisio-saurs to be 30 feet long and it must have been a spectacular sight to

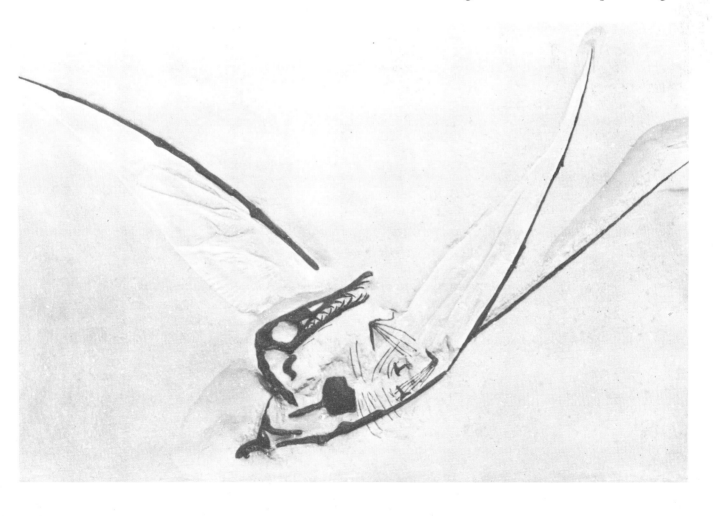

see one of them emerge from the sea and crawl ponderously over the shore to lay their eggs.

In the sea the Pleisiosaurs were accompanied by another group of swimming reptiles called Ichthyosaurs. These were more whale-like in appearance but since they gave birth to their young alive they had no need to return to land to lay eggs.

On land these reptiles created astonishing feats of growth and body design. Carnivorous types like the dinosaur Ceratosaurus had a short bony horn on its snout and ran swiftly on its strong hind limbs, using its three-toed fore-limbs for tearing flesh. Contrasting with these fierce flesh-eating dinosaurs were the enormous types like Brontosaurus, which means 'Thunder Reptile', so-called because of the animal's bulk and weight. Despite its frightening size, it seems to have been a somewhat harmless creature inhabiting the swamps in search of weed-like nourishment. Being 60 feet long and 17 feet high at its hind quarters it must have caused quite a stir as it brushed its way through the mangrove jungles. It shared a similar lifestyle with its larger cousin, the Diplodocus. This type of dinosaur was about 90 feet long, most of which was taken up by a long neck and a very long whip-like tail. Its short head contained exceedingly sharp pointed teeth in front which were used for raking in the plants it needed for nourishment. Another gigantic vegetarian dinosaur was Brachiosaurus which stood 30 feet high and had a heavy elephant-like body and was much more mobile than either the Brontosaurus or Diplodocus.

Some dinosaurs developed a pronounced bi-pedal habit to enable them to move and change direction rapidly. The feet of some, like the Ignanodon, were three-toed while the fore-limbs were five-fingered of which one finger was shaped like a spike. They were often 30 feet long and as far as is known the skin was unprotected by scales. They resemble the modern Iguana lizards, which gave the name to this fossil dinosaur.

Flying Dinosaurs

Almost every experiment in adaptation seems to have been tried by the evolving dinosaurs. One of the most ambitious was to try and take to the air. The most successful was the Pterodactyl. It developed wings spanning about 18 feet which were made of a web of skin stretched from the greatly lengthened fifth finger of the fore-limb to its hind legs. The skull was curiously shaped with a long toothless beak and, as far as flight is concerned it was an aerodynamic freak having such a large, heavy head in relation to the rest of its body. The Ptero-dactyl probably launched itself from cliffs or down steep slopes and glided rather than flew like a bird. However, this first experiment in reptilean flight was apparently not successful as Pterodactyls soon disappeared from the Mesozoic scene. In fact the whole of this ex-travaganza of growth was suddenly brought to a halt and the dino-saurs also vanished from the earth in some catastrophic fashion.

Gryphea incurra – an extinct moving animal which floated in the seas 100 million years ago.

An early submarine : fossil ammonite. This shell fish, now extinct, surfaced by evacuating its many internal chambers indicated by the ornamentation.

Secrets of the Fossils

To unlock the secrets of fossilized remains we must use modern animals and plants as the key; with the vast range of biological information which has been accumulated it is now possible to begin to fill in the gaps concerning the twists and turns which living molecules have taken to initiate a new line of development. This is the most fascinating and astonishing feature of the evolutionary concepts propounded by Darwin as a process of natural selection.

It is surprising what still survives in the post-mortem remains of fossilized animals and plants. This is particularly true of the way in which oxygen has been retained in the shells since the time it became entombed in sediments. Even with the passage of millions of years the shells and skeletons still contain traces of organic matter. The discovery that the oxygen present in the atmosphere is in several forms, called isotopes, has opened up a whole new aspect of the interpretation of life in the past. Of particular interest is the fact that the intake of oxygen isotopes into sea water and its ability to combine with calcium salts to form skeletons and shells is controlled by temperature. From this it has been possible to take a fossil, measure its calcium content and its oxygen isotope content and calculate the temperature of the sea at the time the animal lived.

One of the groups of fossils which has proved especially useful for studying these palaeo-temperatures is the belemnite. These were like cuttle fishes with a bullet-shaped body guard which they used for propulsion or defence. The growth of this solid guard was developed by lime which absorbed the oxygen isotopes from the sea and retained them in proportion to the temperature of the water in which they lived. From the analysis of over 100 specimens of belemnite fossils, Robert Bowen has shown that the temperatures of the seas in Jurassic times (about 150 million years ago) indicated that the earth had larger tropical belts than at present, with the equatorial region running through the U.S.A. and Europe and that the North Pole was situated in the eastern region of Asia as we know it today. Again by calculating the ratio of oxygen isotopes to lime, Robert Bowen has also shown that the North Pole had migrated to somewhere in the region of the Bering Sea by Cretaceous times, that is, about 30 million years later. In this way we shall eventually be able to plot the variations in the temperature of the sea during each geological period and so add yet another facet to the emergent picture of the expanding ocean floors.

The Chemistry of Life

Along with such evidence of primaeval sea temperatures, we are now witnessing the development of one group of animals and plants evolving into new forms of life on a microscopic basis. New methods of analysis are being employed all the time to equip scientists in their hunt for the origin and evolution of life on a molecular basis. Biological cells are now being analyzed in detail and as a result we are

Limestone formed millions of years ago as shell banks on the sea floor.

already in possession of the broad pattern of events which brought life as we know it into being. Perhaps the most significant advance in this biochemical field of knowledge has been the discovery that the organic compounds containing nitrogen, called amino-acids, make up proteins in an orderly fashion. Originally, as mentioned before, these amino-acids, which are the 'seeds' of life, were probably formed thousands of millions of years ago by electrical disturbances like lightning or by the ultra-violet rays of the sun. Even at this primitive stage the organic molecules were arranged into well established patterns each designed to function as organized units in the birth of an animal or a plant-like organism.

Professor Ernest Barrington has shown that among these compounds there are some which have had almost complete control over the development of living processes, even to the degree of determining the sex of an organism. He has also underlined the fact that blood, which has the ability to absorb and utilize oxygen, is itself strongly influenced by these basic groups of compounds some of which have been shown to survive even in trilobites for over 450 million years. The whole fossil record contains traces of these vital amino-acids which played such a central role in the evolution of life at a molecular level.

Among the compounds studied so far on a molecular basis a great deal of information has emerged concerning those called porphyrins. These are molecules of carbon and hydrogen containing nitrogen in which the atoms are arranged as a ring structure. This cyclic formation of atoms endows the molecule with considerable chemical

energy and so enables the porphyrins to occupy key positions in the complex reactions involved in the creation and reproduction of cellular tissue. For example, they have always been the source of respiratory pigments in even the earliest animals and without doubt the source of haemoglobin. In this remarkable substance the 'haem' is a porphyrin complex containing iron and the 'globin' is a protein. From one vertebrate to another the haem remains unchanged but it is and always has been the subtle changes which have taken place in the globin which have resulted in major changes in body behaviour and development. The haemoglobin of the mammalian foetus, for example, differs from the mother in being able to pick up more oxygen – a property required to obtain adequate supplies of oxygen in the uterus. Thus we shall be looking to the porphyrins in substances like haemoglobin for the core and mechanism of each pattern of animal evolution.

Professor Barrington has written that animals 'depend for the co-ordination of their functions upon two great systems. One of these

A brachiopod (Productus) which lived in profusion during Carboniferous times.

The longest living brachiopod Lingula—here fossilized over 600 million years ago—is still a species living today.

is the nervous system, passing coded messages along nerve fibres. The other is the endocrine system, composed of a diversity of endocrine (or ductless) glands which discharge into the blood the chemical substances that we call hormones. These substances act as chemical messengers. Passing around the body in the blood stream, they evoke appropriate adaptive reactions from other organs with which they come into contact'.

He goes on to point out that vertebrates have a number of such hormones typical of which is the hormone insulin produced by the pancreas. Since insulin promotes the use of glucose as a source of energy it will be obvious that all kinds of vertebrates will have their way of life dominated by the secretions of organs such as the pancreas. When inadequate supplies of insulin occur the animal develops diabetes – a disease not exclusive to mankind. Indeed slight differences in the composition of insulin molecules will determine the energy-level of the animal and enable it to develop or fail to develop limbs or protective armour to combat its environment. It could well have been this molecule which stimulated the creation of reptiles as an urge to escape living like fishes or it may have been the substance which caused the demise of the gigantic dinosaurs.

Like the pancreas the pituitary gland, which is situated at the base of the brain in vertebrates, secretes hormones – the two most important of which are arginine vasopressin and oxytocin. These evoke different levels of response in various organs of the body. Arginine vasopressin influences blood pressures and the flow of urine, while oxytocin influences uterine contraction and the release of milk. These are important functions in mammals but not so important in other vertebrates although these two hormones have variants which must have had profound influence over the evolution of the vertebrates as a whole. It is now conceivable that the formation of a new group of hormones such as these may well have caused the origin of mammals from reptiles.

This molecular study of body processes is beginning to yield results. The way in which amino-acids, associated with the mutation of genes, could provide the explanation of the development of various forms of anaemia and throw light on the sudden demise of seemingly well established organisms. The subtle interplay of the hormones suggests a pattern from which fish life was modified to yield reptiles and birds; and it is now possible to say that all these remarkable changes developed at a molecular level. The fantastic bone structures of the dinosaurs, for instance, could well have been due to the excessive development of one hormone or another.

Another important concept which has emerged from the study of living processes is the orderly way in which the amino-acids have arranged themselves in the make-up of proteins. The various ways in which these arrangements were arrived at resulted in the creation of new forms of life. These changes were in many cases a simple substitution of one amino-acid by another. This is reasonably well shown by the evolution of horses, pigs and rabbits from a common ancestor about 80 million years ago. They all developed as a consequence of re-arrangements of the amino-acids in a group of four-footed vegetarians. These changes may have been caused by the plants they chose to eat or some other simple environmental factor.

There is now no limit to which analytical methods will be able to take us through the fossil record. New light will be continuously shed on diseases which afflict animals today by studying their ancestors embedded in sediments. But there are still wide gaps in our understanding of the origin of species. We still have no ancestral knowledge of the trilobites or the graptolites. We do not yet understand the reason why some shell-fish like the brachiopod Lingula seems to have acquired body-processes which could adapt themselves to changing circumstances and yet remain substantially unchanged, in this case, for about 600 million years.

Possibly the most intriguing gap is the point in time and the nature of the circumstances which led to the creation of flowers. Many living fossils, like the coelocanth fish, provide invaluable sources of information and at a molecular level will soon disclose the final proof of Darwin's brilliant concept that all things living on earth arrived by a natural process of selection of the right molecules at the right time and at the right place.

A belemnite – the fossil remains of a bullet-like skeleton of an extinct organism resembling a cuttle fish.

The upper surface of a brachiopod (Productus).

Chapter Six

NEW DISCOVERIES

The Future

Men of science have travelled far since the end of the Stone Age. They have mapped the universe and landed on the moon. They have analysed complex materials and reduced them to their simplest elements. The alchemists' dreams have largely come true. Men have turned trees into newspapers and replaced silk with artificial threads; they have harnessed the atom and tapped the boundless energy in the earth. Today there are no frontiers of knowledge which cannot be assailed; the march of science will produce spectacular results hopefully to usher in an era of social and material affluence.

Often the magnitude of scientific break-throughs create sub-conscious spasms of pessimism so prevalent today. The time has come to re-assess our riches and to find the means of distributing them more widely than before. Although man is basically material-istic, it is to be hoped that he has now grown to appreciate the need for control of his primaeval instincts. Unfortunately, the distribution of raw materials is exceedingly complicated so that the more advanced nations continue to gain command of the less developed areas of the world. Two global wars have accentuated this state of affairs so that want, fear and greed still rule the world and threaten civilization.

If we are honest with ourselves our concern for the survival of animal and plant life stems from the fact that our future depends on them. We are in bondage to the plants even more than to animals because man could survive as a complete vegetarian but not as an absolute carnivore.

Thus the cultivation of plant life will for a very long time mono-polize human labour. Ironically, the most densely populated areas of the world are those which are subject to shortages of food. By the year 2000, it is estimated that the present day population of about 3,400 million will have increased to over 6,000 million. Survival will depend upon agricultural productivity and this in turn depends upon the mineralogical composition of earth and the most important sub-stance of all – water. Who knows, we may ultimately be driven to using the school-boy experiment of burning hydrogen in air to form pure water for certain purposes; certainly desalination of the sea will have to increase on a scale capable of meeting water shortages created by droughts or pollution. What we now regard as water unfit for human consumption will in future have to serve this purpose simply because mankind has set its standards firmly within the realm of materialism. Even in this day and age those who are comfortably off prefer to drink bottled water in countries where the local supply is the

least bit suspect. And it is not too fanciful to predict that bottled drinking water may someday be delivered along with our milk in the morning. This is a feature which probably will develop in the affluent areas of the world.

What then does the future hold for the teeming, proliferating nations outside those relatively restricted areas of the earth? Ease of communication and visual aids will make these countries yearn for the luxuries of so-called civilized standards of life. The result will be that many such countries will decide to proscribe the sophisticated nations from exploiting their natural resources. With our knowledge of the geology of the earth it should now be possible to predict with reasonable accuracy where such situations may arise, and more sens-ibly use such forecasts to create a more equitable utilization of these precious resources.

Minerals from the Sea

All mineral resources on land are rapidly wasting assets. Only the oceans of the world contain the inexhaustible supplies of the in-gredients for the continued support of animal and plant life on earth. We are already extracting minerals from sea water but this will not become our life-supporting reservoir until we have enough cheap sources of energy to make these thermo-chemical processes a practical proposition.

The drive towards the conservation of energy should make avail-able techniques for using sea water to its fullest extent. For island nations and maritime populations the development of desalination power stations is bound to come; and even in the heartlands of the world it will soon be as vital to pump water as it is oil and gas. Today it is a matter of economics but within this century the relative costs of water and oil will have altered, for although over 70% of the earth's surface is covered by water most of it is too heavily charged with dis-solved salts to be of domestic value. At least we can rely upon the sea as a future source of minerals which are being extracted from the earth at an increasing rate. For in each cubic mile of sea water there are about 160 million tons of solids and if these could be extracted and stored they would cover the land surfaces to a depth of about 500 feet. In addition, the erosion of the land is continuously releasing minerals which pass into the sea to form a continuous cycle of replenishment.

So far only four of the 60 important elements dissolved in sea water have been recovered in significant quantities. These are sodium and

Cutting blocks of salt in the dried-out lakes of Ethiopia.

A salt evaporation spiral in Mexico.

chlorine (in the form of common salt), bromine and magnesium. Smaller quantities of other substances are beginning to develop as commercial possibilities. Gathering common salt from the sea is a very ancient practice and today about 6 million tons are produced by solar evaporation of artificial lakes. At the southern end of San Francisco Bay there are 80 square miles of lagoons being evaporated to produce annually over one million tons of salt.

In some countries, such as Sweden and Russia, brine is produced by freezing sea water and this is filtered from the ice and evaporated in kilns to form salt. So the problem of salt supplies is no longer critical – especially when we realize that brines can be produced in enormous quantities from sea water. It will be this very sea water, used to cool the maritime nuclear power stations which will be created all over the world by the close of this century, that will be the source of metals and fertilizing chemicals for the increase in the production of food.

Irrigation from the Sea

A normal plant for the conversion of sea water would produce – at present day cost of water – enough water per day to meet the needs of about 4 million people or irrigate crops over an area of 500 square miles. This would transform life along the north coast of Africa by turning it into the granary of Europe. Even in countries in Europe the use of sea water for cooling purposes could be processed so that the level of salt is reduced to drinkable limits by these plants. Such warm water would be charged with all the trace elements needed to fertilize the fields. This would create horticultural areas of enormous

productivity extending for miles around the power stations. The expense involved in discharging what is virtually boiling water into the sea is considerable. It must be released as far away from the cooling towers as possible since even a small rise in temperature of the intake water means a drop in efficiency of the cooling system. With the use of such hot water for desalination and horticultural purposes, cooling towers could be abolished and with them will go the clouds of steam which pollute the atmosphere and encourage fog. Instead of constructing hideous concrete cooling towers, plastic pipe lines with perforated appendages could be laid to provide farmers with virtually hydroponic water free of charge, and remove all dangers of drought. This desalinated water would contain all the phosphates, nitrates, sulphates and chlorides in solution and would therefore be an inexhaustible source of liquid fertilizer. One has only to look at areas which have risen above sea level in late geological times to see how fertile they are – made so by the trace elements left behind as the sea moved away.

Water Power

The pattern of life on earth will be dominated by supplies of thermal energy coupled with the abolition of waste. At the moment certain forms of thermal energy are not commercially viable but as more conventional sources of heat from solid and liquid fuels are exhausted we shall be forced to develop them regardless of cost. Electricity is the most versatile form of energy man has yet devised and the use of heat to drive dynamos dominates the industrial picture at the moment. The early use of wind to drive windmills and water to turn wheels is

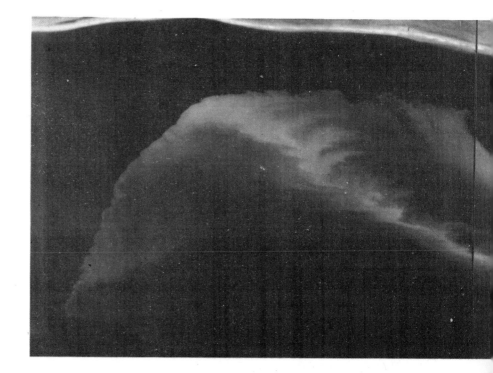

An infra-red aerial photograph of the drift of the warm water outlet at Aberthaw power station in South Wales.

still in basic principle the most elegant means of producing electricity. Capricious though wind action is there are new ideas evolving to create small pockets of energy using revolving drums instead of blades. In certain parts of the world, where wind directions are reasonably constant, this source of energy will come into its own.

Hydro-electricity is being sought with renewed vigour as it combines sources of domestic water supplies with the production of power. And the water is unpolluted by the work it performs in a turbine. The imaginative development of the Tennessee River highlights the basic principles involved: a large volume of water falling a short distance or a small volume descending precipitous slopes. Throughout the 650 miles stretch of the Tennessee River nine dams obstruct the flow as it drops 515 feet. Thus by the construction of relatively small dams vast quantities of cheap, everlasting power have been created.

By its very nature water is copiously produced in mountainous areas where high level dams are reasonably easy to construct. The disadvantage lies in their geographical locations, often considerable distances from large centres of population. Even so they contribute significant amounts of electricity by means of these high heads of water, which in the case of Reisseck in Austria drops 5,800 feet into turbines.

Without doubt, in recent times, the most magnificent use of a head of water has been the huge Kariba Dam across the Zambesi River in Rhodesia. This 400 foot wall of concrete holds back 2,000 square miles of water – the largest artificial lake in the world. From this over 800,000 horse-power is developed from turbines placed in caverns hollowed out in the rock at the foot of the dam. However, it must be remembered that volumes of water of this magnitude are not easy to achieve, but there are multitudes of situations where the flow of rivers is capable of yielding considerable horse-power with dams as low as 30 or 40 feet.

One of the problems of maintaining a power station by means of burning coal, gas or oil is the need to cater for peak demands. Such power stations run efficiently only if the turbines are maintained at a constant speed. This results in excessive outputs of electricity during periods, especially at night, when the demand for power drops. Such demands cannot be met in all cases and neither can unused electricity be stored as such. To meet peak demands, a novel idea is gaining prominence throughout the world. Basically it is to use the off-peak periods of electrical output to perform work which can be used to supplement the output when needed, or when a power station breaks down. One efficient way of doing this is to use surplus electricity to pump water into high level reservoirs or tanks and to release it to drive turbines when peak loads or disasters occur. This is called pumped storage power. About 35,000 megawatts of hydro-electricity is already in service or under construction in Europe, the U.S.A. and Japan. One of the biggest new schemes is one in which an abandoned slate quarry in North Wales is to be used and this will be larger than any similar scheme in Europe. When completed it will be capable of yielding 1400 megawatts which can be released in only 10 seconds from the moment of opening the valves into the turbines.

This pattern of water power is one which will spread internationally and will take on a variety of forms. One idea which will prove

attractive is to use the enormous vacant spaces in abandoned coal mines for storing water and allowing it to fall into deeper abandoned workings through the existing shafts. Such an idea will have many advantages but paramount among these is the coincidence of old coal mines with areas of industry which are afflicted by peak period situations. These ready-made reservoirs with high and low level workings abound in coalfields like South Wales and it is calculated that such water power storage could satisfy the growth needs of its industrial re-birth.

The harnessing of the tides provides another source of electrical energy but this has only been achieved where the discontinuous ebb and flow can be balanced out without serious interruption to shipping – usually by selecting a site where two inlets separated by a neck of land can be joined by dams. So far, only one major tidal-power plant has been brought into full operation and that is in France where the River Rance in Brittany has a tidal rise and fall in the order of 38 feet. This single basin scheme is unique in that it uses turbines which are designed to work on both the rising and falling tides. The future for this method of harnessing natural events is immense and we shall eventually see places like Passamaquoddy, on the eastern seaboard of the U.S.A. or the Bay of Fundy on Canada's eastern coast developed in this way.

Energy from the Sun

The earth receives energy from the Sun amounting to 30,000 times as much as required by mankind today and yet solar energy has been the most neglected to date. Such eternal sources of energy are now uppermost in our minds as the resources of oil and gas are rapidly coming to an end. The sun, the sea and the heat and radioactivity of the earth's interior are the only conceivable sources left to us. Focussing the sun's rays to create intense heat has already begun. The French have covered a mountainside in the Pyrennees with mirrors which are electronically guided to capture sunshine and to focus it on one spot to create a solar furnace. After years of pioneering studies they are now able to create 'hot spots' of over 1500°C which will melt thick metal plates in minutes. From this has been developed the idea that a clean source of heat can be used to purify substances like magnesium oxide required for various industrial purposes.

In a different sphere we see the practical application of the sun rays being converted into electricity in the form of the solar batteries which provide the energy for the satellites travelling in space. These are created by the heat of the sun's rays on layers of silicon metal bonded to a variety of other substances. The heat transfer from one layer to another creates electrical discharges of a continuous character. Riding above the clouds in an orbit of about 22,300 miles radius, these satellites armed with solar cells will remain synchronized with sunshine.

On earth the use of the direct rays of the sun is limited and only in deserts will there be sufficient sunshine to make these cells work with reasonable efficiency. But deserts do not encourage people or industry and this would present objections to the establishment of solar power stations in arid areas. To counter such criticisms research is being done into the possibility of using that fraction of sunshine called daylight. Already houses are being designed like sophisticated greenhouses for the conversion of light into heat.

Windmills of Majorca : one of man's oldest
methods of capturing the power of the wind.

The water wheel is one of man's most elegant
inventions for harnessing Nature's power.

Lake Kariba, Rhodesia, the largest man-made lake in the world harnessed by a dam capable of producing 800,000 horsepower of hydro-electricity.

and metals react and release the electricity created by their inter-action. Working in Cambridge, England over the past quarter-century, F. T. Bacon has succeeded in creating a fuel cell from potassium hydroxide and nickel electrodes. These rods of nickel are filled with minute pores into which the potassium hydroxide soaks its way and causes oxygen and hydrogen to react and release elec-tricity. The Bacon cell has become so successful that it has been used to drive a small car. Cells called 'carbon' fuel cells have been invented which makes use of the chemical reaction of oxygen on carbonaceous fuels such as propane gas. As long ago as 1959 a farm tractor was driven by such a cell for six months.

Eventually ways will be found to use coal as a source of electro-chemical energy – a feat which will revolutionize transport and see the end of the petrol-driven automobiles. No matter which way one looks at the future, it is dominated by an expansion in the use of the material and energy resources of the world. These cannot be separated as increased consumption of materials inevitably demands increasing supplies of energy.

Professor J. O. M. Bockis of Flinders University, South Australia has brought another vision nearer to reality. J. B. S. Haldane sug-gested many years ago that off-peak electricity could be used to pro-duce hydrogen. This gas is no more dangerous than petrol and when burnt would only give off water and not pollute the atmosphere with hydrocarbons. Naturally, this suggestion would not find favour in a petroleum-based economy. Be that as it may, Prof. Bockis and others have shown beyond little doubt that a new world of energy will emerge based on hydrogen. This gas will be produced by fusion reactors visualized as artificial islands in the oceans. From these the hydrogen would be liquified and used to make fuel cells which would replace the combustion engines of cars, trains and ships. These cells would be too heavy in an aircraft such as a Jumbo Jet as they would increase the payload on a Transatlantic flight by 70 tons. So again and again we see scientists responding to a challenge and finding an alternative to the potential disaster which faces the world from the misuse of conventional materials and sources of energy.

Our Mineral Heritage

Looking back on our history the behaviour of mankind does not instil us with much confidence in the future. We always seem hell-bent on self-destruction and yet on the brink of disaster some un-expected turn of events creates a new and hopeful tomorrow. Through hindsight it becomes clear that the more diverse man's knowledge the more he has abused it to the point where doomsday seems to come rushing forward ever faster. When *Homo sapiens* discovered metals he turned them into weapons of war. Swords, daggers and shields made him a more efficient warrior. Nails enabled him to build ships with which he sought to extend his military influence over the seas. Always a materialist, he took advantage of every technological dis-

Quite different from any of these ideas is the highly imaginative and yet practicable possibility of having a huge satellite transmitting microwaves to earth which are collected wherever required and con-verted into electrical current. Many of these solar schemes appear to be too expensive at the moment, but enough evidence exists to show that within the next two or three decades they will become compar-able with the production of electricity as we know it today. It only needs a break-through in the discovery of new materials for capturing sunshine, or even daylight, and converting it either into heat or elec-tricity to make this clean source of energy available at any place on earth.

The New Electricity

Perhaps the most exciting prospect of producing energy is that which aims at producing electricity without dynamoes. There is nothing new in this as the dry battery does it as an electrochemical cell. So far it has only been exploited as a relatively small portable generator since the useful life of these cells is rather limited. It has also proved to be a relatively expensive source of electricity, but this has not deterred scientists from searching for new ways of making chemicals

covery to acquire from others their possessions and to protect his ill-gotten gains. Gun powder revolutionized the art of war and the advent of the machine was immediately hailed as a great military advance.

Endless examples of the misuse of new knowledge can be quoted to show how dangerously man has squandered the products of the earth's crust and its atmosphere. During the past one hundred years man has consumed more of the earth's resources than ever before owing to the immense escalation of the machine age and the recurrence of global wars. As a result there is scarcely a part of the world which has escaped pollution from the waste products of materials extracted from the earth. This, ironically, seems to have brought mankind to its senses and is yet another example of the forces which from time to time dissuades the human race from rushing headlong into irretrievable disaster.

Given time to reassess priorities, the next century should see materialism placed in its proper perspective and replaced by a more beneficent and cooperative way of life. For example, countries like India who now maintain a vast standing army which would be virtually useless if attacked by one of the major nations, should be using some of this labour force to lay drains, create water supplies and so rid the under-nourished millions of epidemics and physical catastrophes.

Some emergent countries containing vast untapped reserves of minerals may try to hold the developed world to ransom by withholding supplies. This is evident in the case of Middle East oil.

The so-called politics of oil has always been an inflammatory subject. Economists have recognized that the world is faced with an energy crisis. This is not because there is an immediate shortage of oil. It is due to the peculiar way in which geological events have created oil-fields in restricted areas of the crust of the earth. Supplies in America are disappearing rapidly and she is now forced to accept dependence upon overseas supplies. So for the first time in her history, the whole basis of her economy will no longer be immune from foreign influences. The U.S.S.R. is still self-supporting, but is surrounded by vast overpopulated territories like China whose progress will depend upon adequate supplies of petroleum.

The present state of our knowledge suggests that it is unlikely that major oilfields remain undiscovered above sea level. Hope lies in the possibility of oil occurring in the sediments which have accumulated on the shelves of the continents. Yet again we see how the knowledge of the crust repeatedly comes to man's rescue and although seemingly academic at first sight, the concept of continental drift by plate tectonics will have a profound effect upon the search for oilfields beneath the sea.

Our new understanding of the mechanism of crustal formation enables one to look afresh at problems of this kind. In the world of tomorrow it will be the countries able to lay claim to submarine oil-

Ffestiniog high level dam in North Wales, storing water used to supply peak period hydro-electricity.

fields which will control the sources of energy and finance. In exactly the same way as the Sheiks of the Middle East have become significant factors in world economy, others will emerge within our century. Fortunately, from our increasing knowledge of the growth and decay of land masses we should be able to prophecy where these areas might be.

The discovery and exploitation of oil and natural gas in the North Sea has opened up the whole concept of exploration of the continental shelves of the world. It is now known that about 60% of the 130,000 square miles of the North Sea contains oil-bearing structures. This compares with the 68,000 square miles of the oilfields of Libya or the 70,000 square miles of Iran. The potential is enormous but so is the cost. Astronomic though the sums of money involved appear,

The solar furnace of the Pyrenees. Myriads of mirrors focus the sun's rays onto an intense hot spot.
The invention of turbines which work on the ebb and flow of the tides has enabled the French to harness the eternal power of the Rance estuary.

the strategic importance for Europe of being able to find an alternative to Middle East oil outweighs everything.

On the creative side this extraction of petroleum and gas from the bowels of the North Sea has produced challenging problems and opportunities for engineers to create hitherto unbelievable structures. The floating drilling rigs are in themselves breathtaking but these will be comparatively simple structures compared with the prefabricated islands of steel which will be built to withstand the storms and tides of the North Sea. The Norwegians have already built such an island. It will store a million barrels of oil abstracted from their share of the Ekofisk oilfield. When the time is right it will be floated from Stavanger harbour like a metallic iceberg rising 40 feet above the waves and descending to a depth of 300 feet when it comes to rest on the sea floor.

Such structures will set the scene for undersea oilfields in many other parts of the world; ultimately we shall have thousands of new islanders developing their own life styles within these vast tanks of oil and gas. As time goes by these islanders will experience the effects of removing enormous volumes of oil and gas from the rocks held under pressures over two miles down by the tectonic plates. These are being pushed against the oil-bearing beds of the continental shelf. The frequent eruption of volcanoes in Iceland is an expression of such submarine forces at work and while there is no connection at the moment between these and the withdrawal of oil, they may in future become significant. Even so, there is very little doubt that exploration northwards in the North Sea for oilfields will become less and less successful as this brittle hot zone of plates is approached. The

An island being constructed by the
Norwegians to be floated out to the Ekofisk
Oilfield in the North Sea.

Pisces, a two man submersible used to work on the ocean bed to depths of up to 3,000 feet.

discovery and exploitation of oil and natural gas in the North Sea is but the prelude to the exciting era of undersea exploration which lies ahead.

In the Irish Sea there are deep channels filled with sediment which undoubtedly contain oil or gas, perhaps both. This 'find' will transform the sources of power available in Eire as the shelf extends many miles westwards into the Atlantic before it encounters the lavas of the plates expanded from the Mid-Atlantic ridge. Likewise, countries such as South Africa, which are clearly not endowed with potential oil-bearing sediments on land, will ultimately discover their own submarine oilfields. Traces of oil have been discovered in the sandstones of Antarctica and these may one day prove sufficient to encourage the oil-men to endure the hazards of drilling at temperatures constantly below zero. Ten years ago under such conditions, the prospect of developing an oilfield in Alaska would have been laughed at, but today it has become one of the most hopeful sources of energy for the U.S.A.

Underwater Vehicles

One of the many 'spin-off' rewards of developing the oil of the North Sea has been the invention of underwater vehicles to go down and inspect pipe lines and perform operations at depths too great for divers. The British Company of Vickers has created 'Pisces' which they claim is five years ahead of everyone else in this field of undersea work. This is a crab-like vehicle which can operate under its own power to a depth of 3,000 feet. Inside this so-called submersible two men can operate remote controlled tools for mechanical work on pipe lines or to excavate trenches. Imagine for a moment the use of such equipment where it is known that large quantities of tin-bearing minerals await recovery. It simply involves the use of a submersible like Pisces to descend to the localities identified by the remote controlled survey car and attach suction pipes to the tin-bearing sands. In this way we shall soon see the actual mining of tin dwarfed by the recovery of these alluvial minerals.

New Resources

The world is running out of mineral resources and the population is exploding beyond the output of food. This is the cry of the voices of doom who refuse to recognize the human capacity to avoid these disasters. It is true that mineral resources now appear to be thin in the ground and equally true that large areas of the world are becoming over populated. Owing to the reckless use of scientific discoveries civilization has created a trail of destruction. But the failure of the past should be the stimulus for attacking the roots of this present situation. When machines began to replace men the short-sighted militants quite naturally tried to resist their development. The invention of nuclear energy has been misused for military purposes and this has aroused widespread revolt; but we should not lose sight of

the fact that nuclear power will ultimately provide mankind with the energy it will need to create more food and new materials for the future. Even the most casual examination of this planet is sufficient to reveal that its very existence depends on the kind of processes which are now regarded as the product of a nuclear power station operating in the earth's interior. We can confidently look ahead to the eventual use of this enormous source of sub-crustal power when all conventional sources have been exhausted.

It used to be thought that the United States possessed mineral wealth to last for centuries. When stock was taken at the end of the last war, it was found that she had consumed 83% of her silver and lead, 78% of her vanadium, 70% of her tungsten, 87% of her mercury, 60% of copper and zinc and all the high grade iron. The same picture emerged in Europe so that today these traditional centres of industry must exploit mineral resources overseas. At the same time consumption is increasing as more machines are made and destroyed in the service of man. Small wonder then that notes of alarm are sounded loudly, but the prophets of doom do not heed the voices of others who are developing processes which will extract even the leanest deposits.

Mounds created by thousands of termites burrowing in the earth and depositing their excreta as a form of iron ore. This mechanism is as yet unknown to man.

Fritz Haber (1869–1934), famous German chemist and Nobel Prize winner who solved the problem of manufacturing ammonia from the atmosphere. He also devised methods of extracting metals like gold from the sea.

Historic trends have always been initiated by discoveries which in their time have passed almost unnoticed. Today we realize that the German chemist Fritz Haber conducted two rather remarkable experiments. He first invented the method of making ammonia from the nitrogen in the air. This freed the world from its previous dependence on nitrates in the ground. Haber was then charged with the task of extracting gold from the sea during the First World War and nearly succeeded. Unfortunately, governments seem to be only interested in such erudite scientific matters when engaged in war. Now is the time to revive his ideas on a large scale. The sea is not only incomparably richer in minerals than mines but it is also constantly replenishing itself from the sediments, and even by man-made effluents, which are constantly being discharged into it. Since the time of Haber chemists have been experimenting with this storehouse of riches.

While gold is relatively unimportant it has always been an attractive constituent of the sea and may eventually be recovered as a by-product of other substances. Already it has been shown by the Dow Company of North Carolina that gold is removed along with the bromine which they extract in commercial quantities. In fact, the oceans of the world have become the main source of bromine which is increasingly required in the manufacture of sophisticated chemicals and plastics. Magnesium is a light metal which is required increasingly for alloys. It used to be obtained from dolomitic limestones or magnesium-rich brines which are not in plentiful supply and expensive to process. The urgent need for magnesium in wartime Britain during the U-boat blockade spurred on her scientists to invent a method of extracting magnesium from sea water. This typical response to necessity was achieved by roasting oyster shells to form milk of lime which, when added to the sea water, converted the magnesium into an insoluble magnesium hydroxide. This compound settled as a sludge. The sludge was then dissolved in hydrochloric acid and placed in an electrolytic cell which deposited the magnesium and released the chlorine to be returned to form fresh acid. This is now a vast continuous process and is the major source of magnesium in the U.S.A. and elsewhere.

Bernard Courtois, a French salt petre manufacturer in 1811, discovered that seaweed contained iodine. He noticed when cleaning his salt petre vats with hot sulphuric acid that a violet vapour arose from the seaweed ash from which the salt petre had been obtained. These violet vapours condensed on the walls of the vessels as dark metallic crystals: these he realized were iodine crystals. Shortly after this, the medical importance of iodine became known and so began the iodine industry. Later the discovery of iodine in the nitrate deposits of Chile put an end to this iodine-from-seaweed industry in Scotland and for nearly a century these seaweeds were neglected. Renewed interest was stimulated during the war in seaweeds as they had been found to contain a remarkable group of substances called

Synthetic diamonds created by subjecting carbon to high temperatures and pressures.

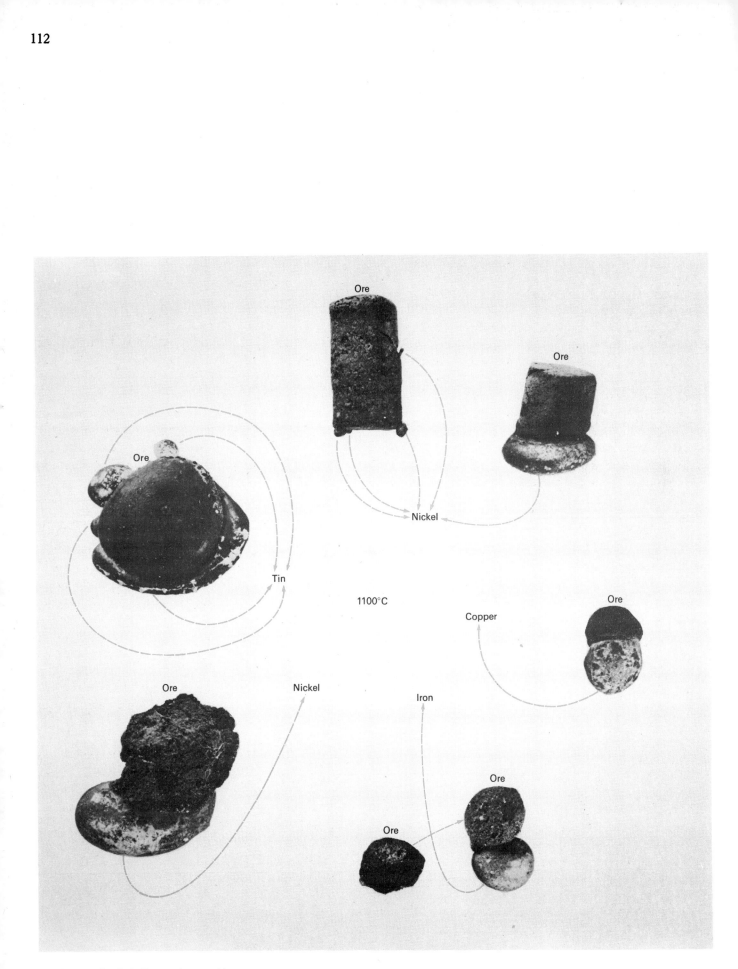

The Contact Catalytic Process invented from
the study of the reactions created when
meteoritic materials enter the earth's
atmosphere. Shown for the first time are
samples of ores which have released their
metals when thrust like meteorites into a high
temperature furnace.

alginates. Although a vast amount of research effort has been expended on the composition of these components of seaweed they are still very much in the realm of tomorrow's materials. There is a substantial industry extracting alginates from seaweed off the coast of Southern California from which the products are mainly used as emulsifying agents in the preparation of foodstuffs. As a gelling agent it is superb and at the same time the sodium and potassium salts of this alginic acid can be spun into threads. When threaded with single strands of cotton, a cloth emerges which can be rendered silk-like in weave by passing it through slightly alkaline waters, which dissolve the alginates. The initially thick cloth is thus rendered single strand in weave and resembles silk. We are only at the beginning of its uses and when the vast seaweed fields of say the Sargasso Sea are cropped and processed a new source of potential proteins, fertilizers and organic fuels will be at our disposal. More significantly, it will be a never ending source of raw materials as seaweed fields are difficult to destroy.

Metals from the Sea

Marine organisms have a capability of concentrating metals in their bodies in a way not yet achieved by chemists. Vanadium is beyond the limits of chemical extraction in sea water, yet tunicates can concentrate it in their mucus by a factor of over 280,000 times its concentration in the water. Fish can take up lead in a phenomenal way. Algae can secrete strontium to such an extent that it is now being looked at as a source of this rare earth metal. These inshore marine organisms are known to contain recoverable quantities of chromium, rubidium, lithium, barium, to mention but a few rare metals which modern technology requires for its sophisticated metallographic and electronic activities. In years to come we shall see whole areas of the sea deliberately encouraged to develop such forms of marine life to provide these rare metals. It will become a form of biological mining of the sea using plankton, algae and seaweeds to carry out the operation and then provide the organic matter from which the metals can be easily extracted.

Mining Plant and Animal Life

Just as the living organisms of the sea digest metals to sustain their living cells so the land plants and animals imbibe the solubilized components of the soil. The common horsetail (Equisetum) has the ability to suck up mineral matter from the soil and deposit it in its cellular walls. This biological fact has already been exploited in Canada where horsetail farms have been developed on mud banks containing traces of gold. Where these plants are cropped and then incinerated their concentrates of gold produce as much of the metal ton for ton as an average gold mine in South Africa. From time to time we shall see this technique repeated in many parts of the world. We shall also see the results of extensive research being carried out on the biological processes which enable plants and animals to render with ease seemingly insoluble substances into solution. Once this is understood a whole new field of mineral extraction will replace the present comparatively crude methods we use today to meet the needs of the metallurgist. Even now we do not fully understand how a common earth worm treats soil and abstracts from it a complement of mineral matter to enable it to carry on its hermaphrodite existence. Similarly the study of the digestive processes of the termite would reap rich rewards. These busy little insects can build earthworks by excreting iron at a rate incapable of being matched by an ironworks. Within their digestive tract they must possess a remarkable group of biological chemicals to perform this feat. This may sound too fanciful to be accepted but it is from the most unlikely sources of scientific knowledge that thoughts like these are confirmed. When building a hut to house some equipment to measure magnetic effects McFadden and Jones of the University of Rhodesia found that their apparatus was made useless by the magnetic field upon which the building stood. Believing that this was due to buried metal they trenched the area and found no metal. What they found was a swarm of termite tunnels which these little beasts had lined with magnetic iron ore. Beyond doubt the termites have created this remnant magnetization by a biological process hitherto unknown to man. Yet another example of the future pattern which will emerge as man begins to learn to use biological energy to create his new magnetic materials without the use of heat. To be able to imitate the biological process of the termite would rid the world of steel works as we know them today and obviate our anxiety concerning the supplies of raw materials and the fuels to smelt them into metals.

Recycling Nature

Concentrates of elements which had previously been regarded as too minute to recover are now fair game for the scientist – even furnace chimneys and sewers are becoming sources of profitable raw materials. Soon we shall reach the desired end-point where everything has a potential use and that the waste product of one enterprise will become the raw materials of another. This is the way to rid the world of pollution. Once again it took a war to open up these possibilities. When Germany was cut off from supplies of sulphur she developed a method of making it from thick deposits of calcium sulphate deposited in geological times as gypsum and anhydrite.

Previously, elementary sulphur was obtained from volcanic areas or the sulphide iron mineral called pyrite (or 'fool's gold') regarded as the most economic sources of this indispensable substance for making sulphuric acid. As a result of the German developments in the last war the widespread deposits of gypsum will no longer be looked upon simply as a source of Plaster of Paris or builder's plaster and plaster boards. It does require a great deal of heat to reduce the gypsum to sulphur and coal was used to supply heat. Recently, experiments are beginning to show that waste materials like sewage can achieve the release of sulphur as sulphur dioxide leaving behind dark clinker which can be used for a variety of building and ceramic purposes; another example of treating natural minerals as chemicals so as to produce not one but at least two valuable end-products.

Lubricants, cloth-fibres, rubber, oils and solvents are derived from natural hydrocarbons. These compounds burn, grow moulds and decay, or they are attacked by rodents, insects and fungi. By adding fluorine to synthetic organic compounds a whole new range of imitations are emerging which offset these disadvantages. The so-called fluoro-carbons do not burn, corrode or decay and they resist attack from rodents, insects and fungi. They will seal engines and provide almost indestructible lubricants and long-life fillers for rubber tyres. But fluorine is not in plentiful supply and until we can extract it from dilute solutions like the sea, the principal source is the mineral called fluorspar.

Fluorspar is a fascinating substance, mainly composed of calcium fluoride. It was the mineral which was found to glow in ultra-violet light and gave the name fluorescence to this phenomenon. This was yet another ordinary discovery which is beginning to occupy the minds of scientists as a source of light, to save electricity. Phosphorescence falls into somewhat the same category as fluorescence and it is predicted that it will soon be commonplace to produce the same effect as a television tube by incorporating phosphorescent substances into paint and wallpaper which will absorb light during the day and radiate it at night. As a consequence minerals will take a new turn in providing such needs. Consequently, there is an intensive interest in the whereabouts of even small quantities of minerals like fluorspar which will thus slowly cease being regarded as a semi-precious stone for jewellery and ornaments, or as a flux for smelting metals. Its future use will be in the production of the most vigorous of all acids – hydrofluoric acid – for making the fluorocarbons of the near future.

Silver has had somewhat the same history. Being a soft bright easily worked metal, its original use was in the manufacture of jewellery, ornaments, utensils and cutlery. It responds easily to electrolysis, so silver plating became important as well as in the manufacture of paints for mirrors. The advent of photography depended upon the response of silver salts to light. As a result, the Eastman Kodak Company soon ranked next to the U.S. Treasury in the consumption of silver. Equally interesting is that the quantity of silver recovered from the perforations in films amounts in weight to more than the output of any silver mine in the world.

Lessons from the Meteorites

The belief that the study of meteorites has no obvious economic value could be further from the truth. They are created under conditions which we cannot simulate in the laboratory. Travelling at speeds of 15 to 50 miles per second they strike the surface of the earth or the moon, and explode to form craters. Study of the fragments has shown that they are composed of materials which range from rocks composed mostly of iron, magnesium and silica to extremely heavy metallic masses mainly consisting of iron and nickel. Associated with these relatively common constituents are some surprising substances. Diamonds have been found in meteorites and this has opened up a new approach to the manufacture of artificial ones. It is now reasonably clear that the explosive forces encountered by the meteorite as it enters the earth's atmosphere provides the mechanism for crystallizing carbon into diamond. Proof came when carbon powder, suspended in a bucket of water, was detonated by an explosive charge transmitted through a block of wood floating on the carbon. Diamonds emerged and sank to the bottom of the bucket.

This detonating effect is not yet completely understood but along with other ways of making artificial diamonds it is reasonably clear that industrial diamonds will be replaced by artificial stones in the next decade. Already the U.S.A. is producing about 6 million carats of artificial diamonds – a fact which has already had a profound influence on the outlook of the diamond merchants of the world. The success of this approach to crystallizing substances like carbon will have interesting ramifications, especially in the production of crystals of silicon metal which also occur in meteorites and nowhere else on earth.

The silicon atom is structurally identical with carbon and as a result compounds can be made which are similar. In other words, the combination of carbon with hydrogen to form hydrocarbons can be matched by similar combinations with silicon to form compounds called silicones. Silicones differ from hydrocarbons in that they can withstand higher temperatures and are much more stable. Without their discovery jet propulsion would not have been possible as normal hydrocarbon lubricants would not stand up to the frictional temperatures involved. The silicone greases provided the answer. Unfortunately silicon is an expensive metal to prepare, but even so it has become a household substance used in the silicone waxes with which furniture and floors are now polished.

Silicon has never been found in earth in a pure form as it readily combines in one way or another with oxygen to form silica or silicates. How it came to be protected from oxidation in the meteorite is a complete mystery. The fact that it still exists suggests the possibility that it can be made artificially by means of meteoritic methods similar to those used for artificial diamonds.

Whoever solves this puzzle will have taken the first step into the 'Silicon Age' when silicones will replace hydrocarbons in the manufacture of man-made fibres and metallic substances harder than steel and yet capable of being treated like plastics. To achieve this oxygen has to be stripped from silica of which there is super-abundant supplies on earth. The products of this successful experiment would

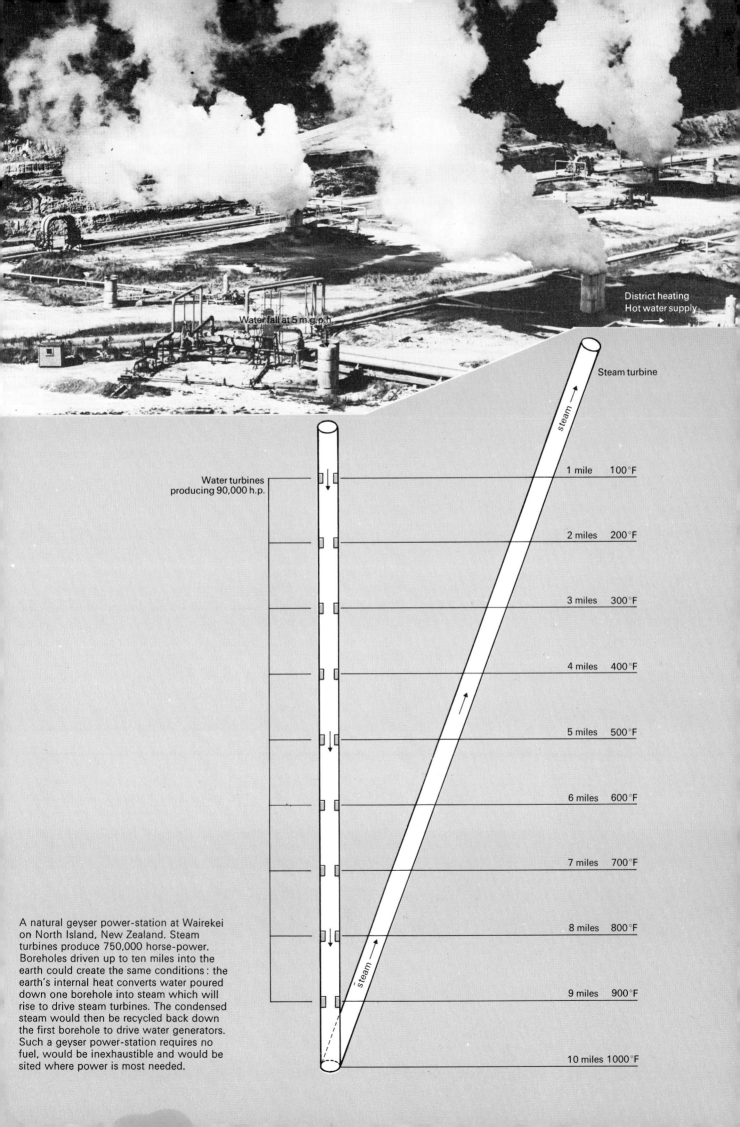

Water fall at 5 m.g.p.h.

District heating
Hot water supply

Steam turbine

steam

Water turbines
producing 90,000 h.p.

steam

steam

1 mile	100°F
2 miles	200°F
3 miles	300°F
4 miles	400°F
5 miles	500°F
6 miles	600°F
7 miles	700°F
8 miles	800°F
9 miles	900°F
10 miles	1000°F

A natural geyser power-station at Wairekei on North Island, New Zealand. Steam turbines produce 750,000 horse-power. Boreholes driven up to ten miles into the earth could create the same conditions: the earth's internal heat converts water poured down one borehole into steam which will rise to drive steam turbines. The condensed steam would then be recycled back down the first borehole to drive water generators. Such a geyser power-station requires no fuel, would be inexhaustible and would be sited where power is most needed.

then be able to forget geological time as a requisite for the formation of stone. Buildings could reappear in stonework as beautiful as those of classical times and we shall see the end of the concrete monstrosities which are disfiguring the townships of the world today.

Another fruitful characteristic of meteorites is their metallic content. Pure nickel can be found along with a host of other metals with which it would normally form alloys. By looking at this singular anomaly, a whole new range of techniques is beginning to emerge for the production of these pure metals from ores now regarded as worthless. The secret behind these metals in meteorites has turned out to be their association with compounds of carbon and hydrogen in the meteorite. At the temperature at which these missiles hit the earth it seems incredible that hydrocarbons could survive, yet mysteriously they do. Large hydrocarbon molecules have been found in meteorites which in the laboratory would have disintegrated at temperatures over 500°C. Yet these meteorites have been raised to temperatures of over 2000°C within the earth's atmosphere.

Scientists have found that if ordinary metallic minerals are mixed with similar hydrocarbons and shot like bullets into high-temperature furnaces they react instantaneously and release metals in a manner similar to meteorites. Electric arc furnaces will at last be fully put to work for the high speed processing of ores which are now regarded as uneconomic. And what is more, this area of investigation has re-opened the role of solid fuels such as coke in metal extraction.

Published here for the first time is a photographic record of the results of some of the experiments carried out to simulate meteoritic conditions. Cylindrical and spherical pellets were made under pressure which contained mixtures of powdered rock and organic matter. These were plunged into an electric furnace at 1100°C – a temperature well below a steel-making furnace. Under the pressures used the organic matter encapsulated the mineral particles. Because of the intense thermal shock the pellets experienced on entering the furnace the organic matter had no time to escape and so reacted instantaneously with the metals in the mineral particles. Organic metallic compounds were formed and these diffused outwards and condensed on the outer surfaces of the pellet. They decomposed as they condensed and so released the pure metal. Thus nickel, which does not melt under 1400°C, was released as an organic vapour condensed as a metal bead on the outside of the cylindrical pellets while they remained intact (see page 112).

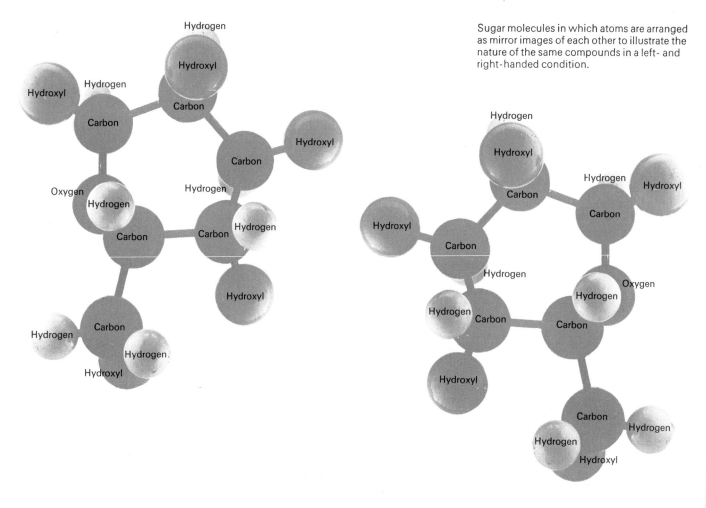

Sugar molecules in which atoms are arranged as mirror images of each other to illustrate the nature of the same compounds in a left- and right-handed condition.

The tiny beads clinging to the surface of the first pellet indicates how even traces of nickel can be recovered from rocks which would be regarded as worthless. The second pellet, with a large attached bead of nickel, is a normal commercial source of nickel, hence the size of the blob of metal. A spherical bead of copper also shows how the metal leaves, in this case, a round pellet of ore.

Iron behaves in the same way. To show that the shape of the pellet has little to do with the process a rough mould of nickel ore has been smelted and this clearly shows how the metal has again been released in a pure condition and the shape of the pellet has remained unchanged. Finally the photograph shows how a spherical pellet of sand has released droplets of tin. This is a process which has been named the 'Contact Catalytic Process' because it is based on bringing organic matter into intimate contact with minerals. The catalytic effect is produced by using raw biogenic substances containing enzymes which trigger off the reactions but having done so plays no further part in the reaction. They are not essential, but when present in the organic substances used they speed up the whole process by acting as catalysts. This novel use of enzymes became important when it was realized that a ready source of these catalysts was town sewage. So the process opened up a new use for one of the most obnoxious products of populated areas. Instead of having sewage farms, towns of the future may well be disposing of their excreta to steel works.

This use of high speed heating and the process of combining organic matter with minerals has extended the use of coal beyond its present narrow limits as a metallurgical fuel. At present only those coals which can be converted into coke can be used for smelting ores, and especially iron ores. Only a limited number of coals will produce coke and there is now an alarming shortage of these in the world today. Japan and Europe are desperately searching the world for new supplies. A replacement for coke is of international importance. This new process may well produce the necessary answer as the experiments have shown that all types of coal, brown coal and even peat can be used instead of coke by means of this 'Contact Catalytic Process'.

Most scientists accept the view that a meteorite resembles the substances which lie beneath the crust and that its iron-nickel content is probably similar to that of the core of this planet. It therefore provides an all-important clue to understanding the nature of the energy locked up in the earth. From the way in which the metals have condensed in meteorites metallurgists can calculate the heat values involved; from actual measurements of rock temperatures recorded in mines and in boreholes there is reasonable evidence to suggest that within conceivable distances into the crust there will be enough heat to convert water into steam. On average the temperature rises 1°C for every 30 meters down. At 20 miles it would be about 500° C. Even at the enormous drilling cost per mile it would be a cheap and ever-lasting source of heat – five times that required to boil water.

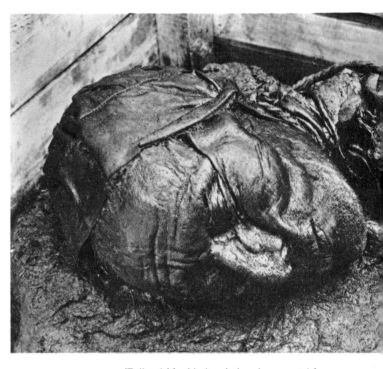

'Tolland Man' beheaded and preserved for 2,000 years in a peat bog in Sweden.

A mammoth leg straight out of the deep freeze. This portion was mummified by permafrost about one million years ago. It was found in Eastern Siberia.

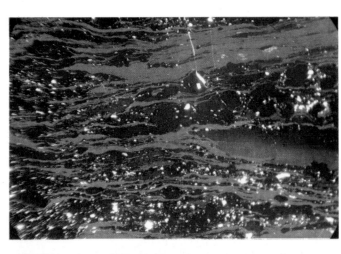

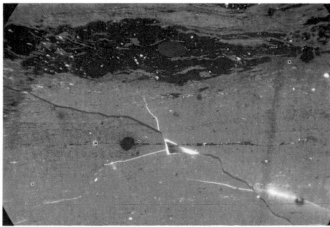

In the sea floor these temperatures would be achieved at much shallower depths and in volcanic areas, even if they are dormant, the boreholes involved are well within the capacity of present day drilling methods. Electricity produced this way would be a highly attractive proposition as it would not pollute the atmosphere with the fumes and steam liberated by a power station or the dangerous use of nuclear reactors.

To tap this thermal energy would, in theory, require two bore-holes drilled at angles so that they would converge and communicate with each other. Turbines fitted into these holes would be driven by water descending in one and by steam generated at the bottom rising like a geyser in the other borehole. The enormous heat of water and steam would not need large volumes of water as it would be recycled. This would be no more costly than the production of nuclear power stations and would be a clean source of heat which would last for ever. It would also be a power station which could be switched off or on at will. Such so-called 'geyser power stations' could be sited where needed and constructed in a relatively short time. When one recalls the rate at which huge shafts have been sunk to a depth of about two miles in places like South Africa the technical problems involved are trivial relative to the construction of solar power stations and the use of nuclear energy.

Life-giving Acids

In an entirely different area of knowledge, meteorites have shed new light on the origin of life itself. The fact that they contain organic molecules has led to speculation that living matter exists somewhere

Transparent sections of coal. The first shows remnants of the wood tissues. The second, the resinous matter produced by decay; and the third is a spore from which these plants grew.
(right) When viewed in polarized light (far right) the spore and other ingredients reveal a spectrum of colours indicating the existence of left-handed and right-handed biological compounds which no longer operate in living tissues today.

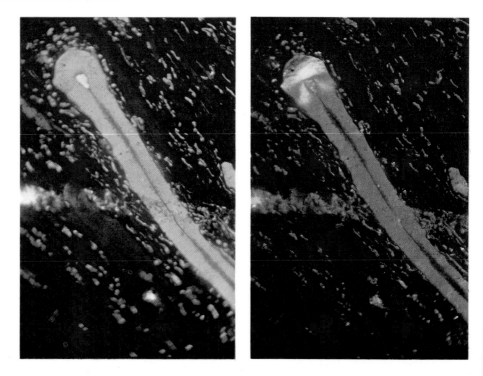

in outer space. Against this is ranged the argument that these organic molecules have been synthesized in the meteorite as it passed through the earth's atmosphere. It is true that the meteorite contains finely divided particles of metals – especially copper and nickel which are widely used by chemists as catalysts to create large organic molecules. Before such organic molecules can be said to be alive they must be further catalyzed into a series of chain reactions. The mechanism for this is only partly understood but the whole process seems to be dominated by the existence of a relatively small group of compounds called amino-acids which are necessary for the formation of proteins. Amino-acids are organic acids in which one hydrogen atom has been replaced by a group of nitrogen compounds.

Of the many potential ways in which this takes place biochemists have found that only 24 of these structures are important in living tissue and only 9 are essential. They have also found that some organisms which are regarded as being simpler than man have greater powers of manufacturing amino-acids. Some, like the red mould on bread (Neurospora), can create them from inorganic constituents. So the route from inanimate materials like rocks to living tissues conceivably progresses by means of micro-organisms such as moulds.

The next surprising feature of the amino-acids is that they will only work if the additional nitrogen compounds are situated on one side of the central acidic molecule. This arises from the fact that amino-acids – sugars and a host of other organic substances – occur in what are commonly called left- or right-handed forms of the same compounds. There is nothing new in this as Biot and Pasteur had shown in the last century that solutions of sugars and tartaric acid caused a beam of polarized light to rotate either to the left or to the right – a movement of light caused by molecules of the same substance having left- or right-handed configurations. Even when these substances were made to crystallize, one crystal was the mirror image of the other. For some reason Nature decided to use only left-handed amino-acids for the production of living matter; but in the case of the sugars both types are employed. Whether this has always been so as regards life on earth has never been resolved; nor was it open to question until right-handed compounds were found in meteorites and also in Carboniferous coals.

Let us suppose that the right-handed compounds in meteorites have been derived from living matter on another exploded planet; then a person from that other world would have evolved in much the same way as earth-man. If he arrived on earth he might be indistinguishable from ourselves but he would not survive. His diet would require right-handed amino-acids and we do not possess them in our natural foods. Consequently, he would die of starvation as he would not be able to digest our proteins.

Thus the existence of right- and left-handed compounds in the early fossils and especially in Carboniferous coals raises important implications and questions. Why, for example, were right-handed

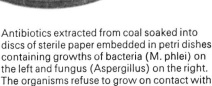

Antibiotics extracted from coal soaked into discs of sterile paper embedded in petri dishes containing growths of bacteria (M. phlei) on the left and fungus (Aspergillus) on the right. The organisms refuse to grow on contact with the antibiotic exuded from the paper discs.

amino-acids abandoned and when did this decision take place? One possible answer is that the changeover took place about 200 million years ago when mammals and flowering plants evolved on earth. Another possibility is that the quantity of nitrogen in the atmosphere altered and caused this changeover to left-handed amino-acids. On balance it would seem more likely that it was the biological changes which accounted for this priority of one type over another. The complicated glandular developments in the mammals may have been responsible for this. In the vegetable world the changeover to processes which created flowering plants may also have been the cause of this selection. One day we shall know and from it may well arise a new understanding of biological processes and particularly those which relate to disease and decay.

The mysterious conditions which have preserved organic matter is involved in these discoveries. Lotus seeds hermetically sealed in vases in China have been taken out after thousands of years and planted to grow trees. Emphasizing the same point is the unique exhibit of the remains of mammoths which have remained in the 'deep freeze' of Eastern Siberia since Stone Age times. The soft tissues and the hairy legs have been perfectly preserved for at least a quarter of a million years. So it would appear that time is not important in the preservation of living tissues – it is the conditions which determines survival.

Biologists at the University of Bradford in England have shown that muds deposited in Lake Windermere 1,500 years ago contain spores which have been cultured back to life. In coal seams deposited over 250 million years ago there are bacteria which live on carbon monoxide. It may be wrong to suggest that these bacteria have survived this enormous period of time buried beneath the earth's surface. On the other hand, if these bacteria have recently entered the coal after it was exposed by the miner why are they not associated with modern forms of bacteria which demand oxygen to live? With this and other factors in mind it was not surprising that these same coals have now been found to contain antibiotics. These are undoubtedly the products of decay which took place over 250 million years ago when the forest debris was attacked and reduced to peat-like deposits by the bacteria and fungi of those times. Since these deposits were developed from flowerless plants they possess unique properties. This is why coals of the so-called Carboniferous age are quite different from coal deposits of newer geological times and contain these products of biogenic decay.

This discovery of antibiotics in coal originated from an appreciation of what happens to a miner when his lungs become filled with coal dust and the extraordinary way in which cuts heal when he is involved in an accident underground. It is difficult to summarize the complicated reactions which take place between the tissues of the lung and the particles of dust, but circumstantial evidence suggests that coal miners are to an extent immune from the effects of tubercular infection and also to the development of lung cancer. The evidence suggests that this immunity is related to the antibiotics found in Carboniferous coals and that these contained both left- and right-handed compounds of biogenic origin. Consequently, as antibiotics they behave in a most unusual way. Normal antibiotics tend to inhibit the growth of certain types of bacteria or fungi. For this reason certain organisms require specific antibiotics. The antibiotics from coal – collectively called 'vitricin' – inhibit the growth of bacteria and fungi alike. This was an unique property and it called for some explanation. In searching for this it was found that coal contained both left- and right-handed forms of the same compounds which suggested that they interacted differently when they came into contact with bacteria or fungi. The different ways in which these two types of disease-producing organisms grow seems to indicate that the right-handed compounds controlled the growth of the fungus by starving it of proteins while the left-handed counterparts resisted the normal growth of the bacteria. This might explain eventually why miners' wounds filled with coal dust resist infection from bacteria and fungi and remain forever dormant as a soft black scar on the body. So we have arrived at the exciting possibility that coal will have a future use in the production of biological chemicals which may be useful in the control of disease in animals and plants.

Crystals and Disease

In so many ways are we finding that the study of the rocks of the earth and those from outer space produces remarkable changes in the outlook of scientists to problems within their specific fields of study. Geology is an inexact science because nature refuses to reveal all her secrets. This is what makes the study of the earth so fascinating. However the development of elegant methods of analysis has enabled the geologist to expand his interests, especially dealing with the origin of life and the factors which tend to destroy or alter the course of living tissues. A good example is how the knowledge of two common rock-minerals, quartz and apatite, have had a profound bearing on the understanding of some diseases. Chemically speaking quartz is silicon dioxide and apatite is calcium phosphate but mineralogically they have the same hexagonal crystal form. Silicon dioxide, which is normally a very insoluble mineral, can be assimulated by normal living processes but when they go wrong they cause crystallization to take place which results in painful end-products in animals and the conversion of plants into stone as exemplified by the petrified forests of Arizona and elsewhere. Calcium phosphate crystallizes in rocks to form a pretty mineral called apatite.

In animals calcium phosphate is a constituent of bones and teeth and any change in body processes causes it to respond often with disastrous results. Possibly the most serious of these processes is the way in which calcium phosphate enters and hardens the middle layer of an artery – the so-called intima. It causes the intima to expand and in reducing the lumen of the artery creates blood pressure and

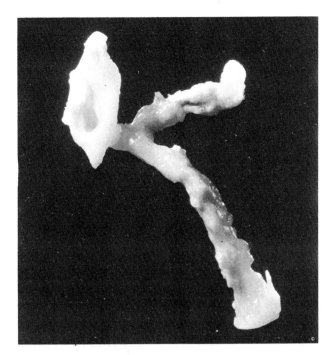

Hardened human arteries emerging from the heart.

A cross-section of a coronary artery showing how the growth of crystalline apatite seals off the artery.

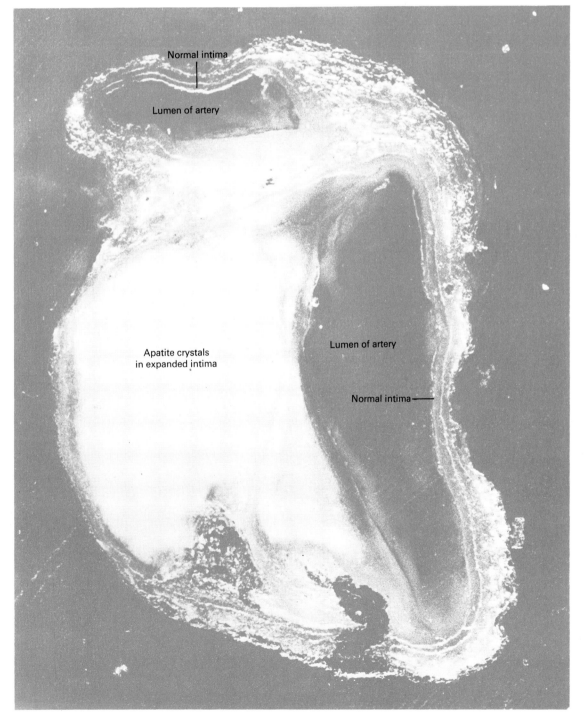

finally coronary thrombosis. Until we treated such arteries as pieces of rock, no-one had seen a section of these arteries as they were at the time of death. The microtomes normally used for cutting sections of arteries were incapable of slicing these hardened ones and so they had to be softened by decalcifying the phosphate. This altered the condition in which the calcium phosphate had developed. Photographs published for the first time (see page 121) show what the artery looked like when the patient suddenly collapsed and died. The clear white area was found to be composed of myriads of crystals of apatite which had formed in the intima causing it to expand and choke the lumen of the artery.

From the pattern of these crystals it was possible to conclude that they had developed by diffusion of solutions of calcium phosphate through the walls of the artery. The next important question was which wall was penetrated. If it was the inner wall then the calcium phosphate had been derived from the blood and it would carry with it other substances, especially iron. No trace of iron was found in these crystals which suggested that the calcium phosphate had diffused through the outer wall from the surrounding tissue. This gave a fresh insight into the cause of coronary thrombosis as it seems to exonerate the quality of the blood.

Since the calcium phosphate appears to come from the surrounding tissue some hope now lies in finding a chemical means of preventing the excessive build-up of calcium salts in the body fluids. The sections of arteries also demanded an explanation of the speed at which these crystals were formed and in some cases redissolved. By normal processes these crystals would be expected to take a long time to form and there is enough evidence to show that coronary thrombosis is a disease associated with age. On the other hand, there is equally strong evidence to show that it was not a disease which causes a person to die as a result of a sudden or sustained exercise as many have died in bed. Even so the final expansion of the intima by apatite crystals is evidently a rapid phenomena. How this takes place is still a mystery. One possible explanation lies in the fact that at nerve endings there are profound electro-chemical forces at work. If these are suddenly stimulated by shock or by fright they could generate enough energy to cause the calcium phosphate to crystallize and solidify the artery. It has been shown in support of this suggestion that among a sample of 400 cases under the age of 61 years most, but not all of these, died with thrombosis from accident or suicide. Gruesome though the subject is it concerns every human being on earth today.

In this odd way the making of rock-like sections of arteries has led us to begin to see some new approaches to one of the most serious afflictions which seems to accompany our rising standards of living. Coronary thrombosis and almost every other form of sclerosis is basically associated with mineralization processes, and it is therefore becoming the concern of the mineralogist who understands how these substances have been formed in the rocks of the earth.

Conclusion

In all kinds of different ways we have seen how the restless earth contains its energy by causing earthquakes and volcanic eruptions. Without these safety valves this planet would have long since exploded and showered the universe with meteorites. We have seen how vast land masses are riding on plates which are constantly being pushed into different positions by ocean floor spreading. With such a powerful interplay of forces we are naturally haunted by the possibility of their spinning out of control. Yet mankind's perennial intellectual exercise of probing the crust has led to discoveries so significant that they have altered the course of one civilization after another. When Madame Curie discovered radium in 1898 who could have predicted then that radioactivity would herald the dawn of the nuclear age 40 years later? From the avalanche of scientific and technological achievement in the field of radioactivity in our own lifetimes we can only hope that this natural source of energy will preserve rather than destroy us.

We have sent men into space to satisfy our curiosity only to find that the view of our earth from spacecraft has opened up entirely new ways of exploring land and sea. To do this we have created visual and audio methods of communication which will ultimately banish ignorance from our world. But for all the great strides forward man has yet to be able to create forces as powerful as those of Nature. The awesome interplay of vast global forces – earthquakes, volcanic eruptions and ocean floor spreading – must temper our view of the future. We now realize that beneath the solid crust of the earth the free-moving particles of radioactive matter create conditions infinitely more powerful than any nuclear power station yet conceived by man. Such knowledge must give us pause for thought when we contemplate sinking boreholes 20 or more miles down towards the centre of the earth. Yet when all the other viable sources of energy have either been exhausted or cannot be extended the storehouse of thermal energy in the crust of the earth will sustain mankind forever.

Ask now the beasts, and they shall teach thee;
and the fowls of the air, and they shall tell thee.
Or speak to the earth and it shall teach thee;
and the fishes of the sea shall declare unto thee.

Book of Job. 12. 7 and 8.

ACKNOWLEDGEMENTS

The Editors gratefully acknowledge the courtesy of the following photographers, artists, publishers, institutions, agencies and corporations for the illustrations in this volume.

Cover
Ernst Hass/John Hillelson Agency
Title page
A. C. Waltham
p. 6 The Mansell Collection
p. 8/9 Victoria and Albert Museum: Crown Copyright
p. 10 Lord Energlyn
p. 11 C. M. Dixon
J. Allan Cash
p. 12 Bruce Coleman Ltd.
p. 13 Crown Copyright reserved: Ministry of Environment
Wendy Smith
p. 14 Eileen Deste
p. 15 By courtesy of the Trustees of the British Museum
p. 16/17 The Mansell Collection
p. 18/19 A. C. Waltham
A. C. Waltham
p. 21 De Beers Consolidated Mines Ltd.
p. 22/23 De Beers Consolidated Mines Ltd.
p. 24 De Beers Consolidated Mines Ltd.
p. 26 Lord Energlyn
Camera Press
p. 27 Novosti
p. 28/29 De Beers Consolidated Mines Ltd.
p. 30 Camera Press
p. 31 De Beers Consolidated Mines Ltd.
De Beers Consolidated Mines Ltd.
Lord Energlyn
p. 32 Eileen Deste
Camera Press
p. 33 Novosti
p. 34 Ronan Picture Library
Ronan Picture Library
p. 35 The British Petroleum Company Limited
The British Petroleum Company Limited
p. 36 British Steel Corporation
p. 37 Japan Information Centre
p. 38 Christopher Marshall
p. 39 Camera Press
p. 40/41 Camera Press
p. 41 Camera Press
p. 42 Institute of Geological Sciences/A. Lacroix
Geoffrey Watkinson
p. 43 ZFA
ZFA
p. 45 Astor Magnusson/Camera Press
p. 46 Lord Energlyn
p. 47 Camera Press
p. 48 J. Allan Cash
p. 48/49 Eileen Deste
p. 49 Lord Energlyn
Lord Energlyn
Lord Energlyn
p. 50 Lord Energlyn
Lord Energlyn
p. 50/51 Lord Energlyn
p. 52 Gerald Cubitt/Camera Press
Lord Energlyn
Lord Energlyn
Lord Energlyn
p. 53 Lamont – Doherty Geological Observatory
Nick Jackson
p. 54 Iclandic Embassy/Herman Schlenker
Iclandic Embassy/Herman Schlenker
p. 55 Lord Energlyn
Lord Energlyn
p. 57 Camera Press/NASA

p. 58 Camera Press/NASA
p. 59 Georg Gerster/John Hillelson Agency Ltd.
p. 60 Camera Press/NASA
p. 62 Christopher Marshall
p. 63 A–Z Botanical Collection
A–Z Botanical Collection
p. 64 Tom McArthur
p. 65 Tom McArthur
p. 66 John Smith
p. 67 John Smith
p. 68 Georg Gerster/John Hillelson Agency Ltd.
p. 69 J. Allan Cash
p. 70/71 John Smith
p. 72 Georg Gerster/John Hillelson Agency Ltd.
p. 73 Black Star
J. P. Charbonnier/Camera Press
p. 74 Christopher Marshall
p. 75 David Nockels
p. 77 Geoffrey Watkinson
p. 78 Geology Department, University College, London
Geology Department, University College, London
Lord Energlyn
p. 79 A. C. Waltham
Lord Energlyn
p. 80 Geology Department, University College, London
p. 81 A. C. Waltham
p. 82 Geology Department, University College, London
p. 84 Geology Department, University College, London
Geology Department, University College, London
p. 85 J. Allan Cash
p. 86/87 A. C. Waltham
p. 88 By courtesy of the Trustees of the British Museum
By courtesy of the Trustees of the British Museum
p. 89 The Mansell Collection
p. 90 A. C. Waltham
p. 91 A. C. Waltham
p. 92/93 Geology Department, University College, London
p. 93 A. C. Waltham
p. 95 J. Allan Cash
p. 96/97 Geology Department, University College, London
Geology Department, University College, London
Geology Department, University College, London
p. 99 Georg Gerster/John Hillelson Agency Ltd.
Georg Gerster/John Hillelson Agency Ltd.
p. 100/101 Crown Copyright Reserved: C.O.I.
p. 102 Camera Press
p. 103 J. Allan Cash
p. 104 Camera Press
p. 105 Central London Electricity Generating Board
p. 106 Aerofilms © Aldus Books Ltd., London
p. 106/107 André Steiner
p. 108 Petroleum Times
p. 109 Vickers Ltd.
p. 110 Bruce Coleman Ltd.
p. 111 Popperfoto
De Beers Consolidated Mines Ltd.
p. 112 Lord Energlyn
p. 115 Camera Press

p. 116 Lord Energlyn
p. 117 Gylendal
Novosti
p. 118 Lord Energlyn
Lord Energlyn
Lord Energlyn
Lord Energlyn
p. 119 Lord Energlyn
p. 121 Lord Energlyn
Lord Energlyn
p. 123 Ernst Hass/John Hillelson Agency Ltd.